Scientific Programmer's Toolkit

Turbo Pascal Edition

SCIENTIFIC PROGRAMMER'S TOOLKIT

Turbo Pascal Edition

M H Beilby

School of Mathematics and Statistics,
University of Birmingham

R D Harding and M R Manning

Department of Applied Mathematics and Theoretical Physics,
University of Cambridge

Adam Hilger
Bristol, Philadelphia and New York

British Library Cataloguing in Publication Data

Beilby, M. H. (Michael Harry) *1943–*
   Scientific programmer's toolkit.
   1. Computer systems. Programming languages
   I. Title   II. Harding, Robert D.   III. Manning, M. R.
   005.133

ISBN 978-0-7503-0127-5

Library of Congress Cataloging-in-Publication Data are available

Consultant Editor: **R D Harding**, University of Cambridge

Published under the Adam Hilger imprint by IOP Publishing Ltd
Techno House, Redcliffe Way, Bristol BS1 6NX, England
335 East 45th Street, New York, NY 10017-3483, USA

US Editorial Office: 1411 Walnut Street, Philadelphia, PA 19102

Printed in Great Britain by J W Arrowsmith Ltd, Bristol

⟩ Contents

⟩ Acknowledgments

We would like to thank Stewart White for his help with the screen drivers used by this Toolkit.

Many of the mathematical routines have been adapted from *Computer Methods for Mathematical Computations* by Forsythe, Malcolm and Moler (Prentice-Hall, 1977).

⟩ Guide to installing the software

You can, if you wish, use the discs supplied with this book directly in your computer. We would not recommend this, however: the discs do not have an infinite life, and will wear out in time. More importantly, it is much too easy to erase the contents of a disc completely; if you erased one of the Toolkit discs, you would have to purchase another copy.

So we strongly advise you to *install* the Toolkit programs you have bought by copying them onto other discs. To make this process easier, each Toolkit disc contains a program called INSTALL, which performs the copying for you.

You can choose to copy the Toolkit discs onto other floppy discs; and, if your machine does not have a hard disc drive, you will have no option other than to do this. However, if your machine does have a hard disc drive, you would be wise to copy the Toolkit discs onto it. A hard disc is both much larger, and very much faster, than any floppy. This means that you can develop more (and larger) programs, more quickly with a hard disc than you can with a floppy disc.

⟩ 1   Hard disc systems

The Toolkit programs may come on one, or several, floppy discs, depending on the type of floppy disc, and whether you have ordered various optional extras (such as the source of the Toolkit). Each disc, however, comes complete with a copy of the INSTALL program.

To copy all the files from one Toolkit disc onto your hard disc, first place the Toolkit disc into floppy disc drive A: of your computer, and issue the DOS command:

    A:

This makes the drive containing the Toolkit files the current drive.

We shall assume that your hard disc drive is drive C: (which is most likely); if it has some other name, you should replace C: by the appropriate drive name. (For example, if your hard disc drive is drive D:, you should use D: rather than C: throughout the following.)

By default, the INSTALL program places the Toolkit code into the directory \TOOLKIT. Thus issuing the DOS command:

    INSTALL C:

will result in the INSTALL program copying all the files from the Toolkit floppy disc to the directory \TOOLKIT on C:, your hard disc drive. If the directory \TOOLKIT does not exist, INSTALL will create it before starting to copy the files.

If you would prefer to install the Toolkit files to some other directory, merely give that directory's name after the drive name. For example, to install the Toolkit files into the directory \MYDIR, say, use the DOS command:

    INSTALL C:\MYDIR

You should repeat this procedure for each Toolkit disc. At the end of this process, the directory \TOOLKIT (or whichever directory you have chosen to contain the Toolkit files) will contain all the Toolkit files from every disc. This is another advantage of using a hard disc to hold the Toolkit files: they will all fit onto it!

⟩ 2   Double-drive floppy disc systems

If your computer has two floppy disc drives, but no hard disc drive, you should install the Toolkit onto new, formatted floppy discs. Throughout this, we shall assume that drive A: on your computer contains the system and DOS files, and use drive B: to contain the installed copies of the Toolkit.

You will need as many new discs as you have Toolkit discs. To format the new discs, issue the DOS command:

    FORMAT B:

Your computer will produce a message like:

        Insert new diskette for drive B:
        and strike ENTER when ready

and you should do as it says: take one of the new discs, and place it in drive B: of your computer (the *other* floppy drive, not the one containing your system disc). When you have done this, and closed the drive door, press `Enter` to start the formatting process. Continue in this way until you have formatted all the new discs.

Then remove your system disc from drive A:, and place one of the Toolkit discs into it. Issue the command:

        A:

to make the drive containing the Toolkit files your current drive. Check that one of the new discs is in drive B:, and issue the command:

    INSTALL B:

This will copy all the files on the Toolkit disc into the root directory of the new floppy disc. If you wish to copy the Toolkit files to some other directory, type the directory name immediately after the drive name. For example, to install the files from the Toolkit disc in the directory **\MYDIR** on drive B:, use the command:

    INSTALL B:\MYDIR

You should then repeat this installation procedure for each Toolkit disc you wish to copy.

One word of warning: the **INSTALL** program will not be able to work if the capacity of drive B: is less than that of drive A: (for example, drive A: is a high-density drive, but drive B: is not). The reason for this is that Toolkit discs are supplied fairly full, and there is not enough room on a lower-capacity disc to hold all the files from a high-capacity disc. If you do have disc drives with different capacities, either install from the lower-capacity drive to the higher-capacity one; or use the procedure recommended for single-drive machines.

⟩ 3   Single-drive floppy disc systems

On a single-drive machine, DOS still behaves as though you had two floppy disc drives, A: and B:, and issues messages reminding you to change the disc where appropriate. For example, assume that you had the disc for drive A: in the floppy drive, and DOS wished to use drive B:; it would issue the message:

```
Insert diskette for drive B: and strike
any key when ready
```

to prompt you to remove the disc for drive A:, and insert that for B:.

You can, therefore, treat a single-drive machine as though it had two drives, A: and B:, and follow the instructions for installation on a double-drive floppy disc system.

However, you will get so many prompts asking you to change between the discs for drives A: and B: that the whole process will become very tedious. The simplest thing to do is to format new discs, one for each Toolkit disc, as for the double-drive system. Then, for each Toolkit disc, issue the DOS command:

```
DISKCOPY A: B:
```

This will issue the message:

```
Insert SOURCE diskette in drive A:

Press any key when ready . . .
```

When you see this, you should remove your system disc, and replace it with the Toolkit disc you wish to copy; then press a key. DISKCOPY will then read the entire disc into memory, and issue the message:

```
Insert TARGET diskette in drive A:

Press any key when ready . . .
```

When you see this, remove the Toolkit disc, and insert one of the formatted new discs. DISKCOPY will then copy all the files from the Toolkit onto the new floppy disc.

Using DISKCOPY means that you only need to change the discs once, rather than once for each file (as you would have to using INSTALL). The only disadvantage of using DISKCOPY is that you cannot decide the directory to hold the Toolkit files: they will be copied onto the root directory of the new floppy disc. This is not normally a problem.

Be warned that you should only use DISKCOPY to copy files onto newly formatted or blank discs; DISKCOPY will delete any files on the target disc.

⟩ 4   Network systems

If your computer is attached to a network, you should install the Toolkit files onto the network disc (which will be almost as fast, and certainly at least as large, as a hard disc drive on your own machine).

The **INSTALL** program treats network discs as identical to hard disc drives; you should refer to Section 1 of this guide, which discusses installation on hard disc drive systems, for details.

⟩ 5   Reformatting compressed source

If you have purchased the source of the Toolkit, you will find that each of the Pascal files has been processed to save disc space. The files still contain legal Pascal, and Turbo Pascal will happily compile them, but you will find them rather unreadable yourself, because many spaces and ends-of-lines have been removed.

To compensate for this, we include with these discs a copy of a Pascal reformatting program called **REFORM**, which restores the compressed files to their original, human-readable format. To reformat the file **CIT_PRIM.PAS**, for example, issue the DOS command:

```
REFORM CIT_PRIM.PAS
```

REFORM will supply the **.PAS** extension to a file name if necessary, so the command:

```
REFORM CIT_PRIM
```

will have the same effect.

REFORM displays a count of the line number as it processes the file; after it has finished, you will find that **CIT_PRIM.PAS** contains a much more readable version of the **Cit_prim** unit.

You can repeat this process with all the other Toolkit units, but be warned that you may not have enough disc space to hold all the units in uncompressed format (unless you have a hard disc, or a high-capacity floppy disc drive).

You can also use **REFORM** to reformat your own Pascal programs into a standard format if you wish. The file **REFORM.DOC** contains brief details of the changes made by **REFORM**, and a list of its options.

⟩ Guide to running the software

There are two types of program on the Toolkit discs: programs which can be executed at once (present as .EXE files, which are ready for execution), and programs which need to be compiled before they can be run (present as .PAS files, which contain Pascal programming language). We deal with these two categories in turn.

⟩ 1   Executable programs

The discs supplied with the *Scientific Programmer's Toolkit* contain many ready-to-run programs. You need no software beyond that provided on the Toolkit discs to run these, not even a Pascal compiler, and the programs require a minimum configuration PC-compatible as long as it has a graphics adapter.

We assume that you have some familiarity with DOS. Your DOS manual will give full details of operating system commands and conventions. You will recall (or the DOS manual will tell you) that the DIR command can be used with a mask; this restricts its output to selected files only. For example, to get a listing of files with extension .EXE you could type:

        DIR *.EXE/W

(The /W makes the DIR command display the file names across the width of the screen.) Any file whose name has the extension .EXE can be run simply by typing its name after the standard DOS prompt. (You do not need to type the .EXE.) For example, if the file PROGRAM.EXE were in your current directory, you could run it by typing either:

        PROGRAM.EXE

or (omitting the .EXE):

        PROGRAM

⟩ *1.1*   *Hard disc systems*

Start up your system in the usual way. If the *Scientific Programmer's Toolkit* discs have not yet been installed, install them now by following the instructions in the *Guide to installing the software* earlier in this book.

You should now find the directory containing the file you wish to run, and make it the current directory. Then type the file's name, ending with Enter (or Return on some machines). For example, the Toolkit files are installed on a hard disc system, by default, in the directory \TOOLKIT. So the following command will run the demonstration programs:

```
CD \TOOLKIT\DEMO
MENU
```

⟩ *1.2*   *Floppy disc systems*

Start up your system in the usual way and decide whether you wish to use drive A: or B: for the Toolkit disc. Make that drive the current drive, find the directory containing the file you wish to run, and make it the current directory. (Note that, by default, the INSTALL program copies Toolkit files into the root directory of a floppy disc; hence demonstration programs will be under the directory \DEMO, and so on.)

Finally, type the program name, ending with Enter (or Return on some machines). As an example of all this, consider mounting a Toolkit disc on drive B:, and running the program MENU in the directory \DEMO. The sequence of DOS commands to do this is the following:

```
B:
CD \DEMO
MENU
```

⟩ **2**   **Programs provided in source code**

Programs provided in source code have file name extensions .PAS. To run these, you will need a Pascal compiler, and we recommend that you use Turbo Pascal version 4.0, or a later release. With Turbo Pascal it is a very straightforward matter to run these programs, especially if you have a hard disc. If you have a Pascal compiler other than Turbo Pascal then you will probably have to adapt the Level 0 routines: Appendices 1 and 2 provide the details. In the rest of this section we will assume that you are using Turbo Pascal version 4.0, 5.0, 5.5, or higher.

Whatever combination of hard disc or floppy discs you are using, the Turbo Pascal compiler must be told where to find all the files it needs to carry out the compilation, and where to put any it creates. By default the current directory is used for everything, but most users prefer to put files into particular directories, according to their purpose. In the main menu (which comes up when you enter Turbo Pascal) there is an **Option** menu which can be used to tell Turbo Pascal which directories to use, and if you are not familiar with it we recommend that you consult your Turbo Pascal manual for the details. Both Turbo Pascal itself and the *Scientific Programmer's Toolkit* make extensive use of *units*, which are libraries of pre-compiled routines. This is not the place to give a full discussion of units (see Chapter 1), but in order to understand how to compile and run programs you need to know that the Toolkit uses two distinct groups of units: the Turbo Pascal system units (such as **Crt**, **Graph**, *etc.*), and Toolkit units. Toolkit units have names beginning with **Cit_**; for example, **Cit_core**, **Cit_prim**, **Cit_grap**, *etc.* Compiled code for these units are stored in files with the extension **.TPU**: for example, the file **CIT_CORE.TPU** holds the code for the unit **Cit_core**, and so on.

⟩ 2.1   *Hard disc systems*

The objective now is to set up Turbo Pascal so that it finds both the Toolkit units (and its own internal units) on your disc. If you have installed the *Scientific Programmer's Toolkit* software using the **INSTALL** program (as described in the *Guide to installing the software*), the Toolkit units will normally be kept in the directory **C:\TOOLKIT\UNITS**. We will suppose that the Turbo Pascal system units are stored in the directory **C:\TURBO**.

To make Turbo Pascal recognise the location of these files, you should call up Turbo Pascal, and press 〇 to select the **Options** menu. Now press 𝔇 to select the **Directories** sub-menu, and set the following directories:

```
Turbo    directory:   C:\TURBO
Object directories:   C:\TOOLKIT\UNITS
Unit    directories:   C:\TURBO;C:\TOOLKIT\UNITS
```

That is, you should ensure that:

- The **Turbo** directory contains your copy of Turbo Pascal.
- The **Object** directories include the one containing the Toolkit code.
- The **Unit** directories include *both* the directory containing Turbo Pascal, and the directory containing Toolkit code.

You may also wish to add other directories at the same time.

Note that when Turbo Pascal asks for **directories** (plural) you can supply a path (that is, a list of directories separated by semicolons). This list must contain the directories shown above, in addition to any other directories you may wish to include.

Next, select the **Compiler** sub-menu. There are two settings here that you will need to make according to whether you are using a hardware floating-point processor (e.g. an 8087) or not. We recommend either:

```
Numeric processing     Software
Conditional defines    REAL
```

or (if your machines has an 8087 or 80287 maths coprocessor):

```
Numeric processing     Hardware
Conditional defines    DOUBLE
```

Note that the choice of **Hardware** or **Software** must correspond to the set of **CIT_*.TPU** files you chose when ordering the *Scientific Programmer's Toolkit*: see the disc label if in doubt. Note also that the conditional defines mentioned above only come into effect when the Toolkit code is recompiled. You will not be able to do this unless you also have the *Units Source Disc*. This is an option, not included in the standard *Scientific Programmer's Toolkit*.

We mention the above conditional defines for completeness; they are irrelevant if you will not be recompiling the units. Note that you can also specify the conditional define **CHECKING**, which makes the evaluator test for numeric overflow; this is described more fully in Chapter 9.

Another important option in the **Compiler** sub-menu is **Link buffer**. This is usually set to **memory**, which is the quickest option for the link stage (i.e. the pause after compilation is complete whilst pre-compiled units and library code are merged). But with large programs it is possible that an error such as "**out of memory**" may occur; you should then try changing **Link buffer** to **disk**. Whilst we are talking about memory problems, you might like to note that setting the **Destination** option in the **Compile** menu to **disk** will achieve a further saving of memory space.

To save having to set these options each time you run Turbo Pascal, you can save them on disc by using the **Options** sub-menu **Save Options** (or, alternatively, use the program **TINST** to build them into the Turbo Pascal compiler itself).

You must also ensure that an appropriate graphics driver is copied into the current directory (if compiling to memory) or in the same executable directory (if compiling to disc). Graphics drivers are in files with extension .**BGI** and are supplied with Turbo Pascal itself. The most commonly required drivers are **CGA.BGI** and **EGAVGA.BGI**.

You are now ready to compile and run a Toolkit program. For your first attempt, we suggest that you make **C:\TOOLKIT\EXAMPLES** the current directory either before or after calling Turbo Pascal (according to your usual practice, and to where you want to keep your Turbo Pascal configuration (*.TP) and pick (*.PCK) files).

Then try **CHAP4\1-1BISEC.PAS**: this is a good test to start with because it does not use any units. To load and run it, select Turbo Pascal's **File** menu, and press $\boxed{\text{L}}$ to invoke the **Load** sub-option. Now type in the file name **CHAP4\1-1BISEC.PAS**, and use the **Run** option ($\boxed{\text{Alt}}$/$\boxed{\text{R}}$ is the quickest way of doing this).

Then try **CHAP4\1-3ROOTS.PAS**, which does use units. After that, browse as you wish.

⟩ 2.2   *Floppy disc systems*

DOS systems always allow for two 'logical' disc drives **A:** and **B:** even if only one physical drive is present. The procedure for running Toolkit source files is therefore the same whether you have one or two floppy disc drives, except that if you have a single drive, the system will prompt you from time to time to change or insert discs.

Using any high-level language on any computer with a single disc drive of low capacity is always likely to be cumbersome, and anyone undertaking extensive programming on such a system is sure to want to upgrade it sooner or later. Nevertheless, we have made sure that every example program in the *Scientific Programmer's Toolkit* can in fact be run on a single-drive system.

To prepare for compilation on one- or two-drive systems, we advise making up a set of working discs from your working master discs. There

is no unique way of setting these up and you can vary the following
suggestion according to your preference. We are assuming that you
have drives of 360K capacity: if you have a higher-capacity drive then
you may be able to place all the files mentioned below on a single disc.

We suggest a *system/Turbo disc* and a *working disc*. The system disc
could have on it DOS and essential files from your Turbo Pascal master
disc. To make a disc with DOS on it you format a disc using the /S
option: for example:

```
FORMAT B:/S
```

This also copies the file COMMAND.COM onto the disc. You must then
copy or create files CONFIG.SYS and AUTOEXEC.BAT according to your
usual requirements. Now copy the essential Turbo files from your Turbo
Pascal master disc: for example:

```
COPY A:TURBO.* B:
...
COPY A:*.TPU B:
...
```

That completes your system/Turbo disc, which you will normally use in
drive A:. To create your working disc:
1. Take a blank disc and format it *without* copying the DOS system,
   then
2. Use INSTALL to copy over the Toolkit units onto the new disc.
3. Now copy whatever .BGI graphics driver you need from your master
   Turbo Pascal disc, and
4. Copy whatever source files you wish to compile.
For example:

```
FORMAT B:
...


< insert master Toolkit Units disc in A: >
A:
INSTALL B:
...
```

```
< insert Turbo Pascal master disc in A: >
COPY A:EGAVGA.BGI B:
COPY A:CGA.BGI B:
...


< insert Toolkit Examples disc in A: >
COPY A:\EXAMPLES\CHAP4\1-1BISEC.PAS B:
COPY A:\EXAMPLES\CHAP4\1-3ROOTS.PAS B:
...
```

We now have to set up Turbo Pascal so that it finds both the Toolkit units (and its own internal units) on the appropriate disc. If you have installed the *Scientific Programmer's Toolkit* software using the **INSTALL** program (as described in the *Guide to installing the software*), the Toolkit units will normally be kept in the root directory of your new disc (mounted on drive **B:**). We will suppose that the Turbo Pascal system units are stored in the root directory of drive **A:**.

To make Turbo Pascal recognise the location of these files, you should call up Turbo Pascal, and press ☐O to select the **Options** menu. Now press ☐D to select the **Directories** sub-menu, and set the following directories:

```
Turbo     directory:  A:\
Object directories:  B:\UNITS
Unit    directories:  A:\;B:\UNITS
```

That is, you should ensure that:

- The **Turbo** directory contains your copy of Turbo Pascal.
- The **Object** directories include the one containing the Toolkit code.
- The **Unit** directories include *both* the directory containing Turbo Pascal, and the directory containing Toolkit code.

You may also wish to add other directories at the same time.

Note that when Turbo Pascal asks for **directories** (plural) you can supply a path (that is, a list of directories separated by semicolons). This list must contain the directories shown above, in addition to any other directories you may wish to include.

Next, select the `Compiler` sub-menu. There are two settings here that you will need to make according to whether you are using a hardware floating-point processor (e.g. an 8087) or not. We recommend either:

```
Numeric processing      Software
Conditional defines     REAL
```

or (if your machines has an 8087 or 80287 maths coprocessor):

```
Numeric processing      Hardware
Conditional defines     DOUBLE
```

Note that the choice of `Hardware` or `Software` must correspond to the set of `CIT_*.TPU` files you chose when ordering the *Scientific Programmer's Toolkit*: see the disc label if in doubt. Note also that the conditional defines mentioned above only come into effect when the Toolkit code is recompiled. You will not be able to do this unless you also have the *Units Source Disc*. This is an option, not included in the standard *Scientific Programmer's Toolkit*.

We mention the above conditional defines for completeness; they are irrelevant if you will not be recompiling the units. Note that you can also specify the conditional define `CHECKING`, which makes the evaluator test for numeric overflow; this is described more fully in Chapter 9.

Another important option in the `Compiler` sub-menu is `Link buffer`. This is usually set to `memory` which is the quickest option for the link stage (i.e. the pause after compilation is complete whilst pre-compiled units and library code are merged). But with large programs it is possible that an error such as "out of memory" may occur; you should then try changing `Link buffer` to `disk`. Whilst we are talking about memory problems, you might like to note that setting the `Destination` option in the `Compile` menu to `disk` will achieve a further economy of memory space. Please note that even these actions may not get you out of trouble because disc space may also be in short supply. If this happens, then the time has come to recognise that your program has outgrown a floppy disc system and you'll need a hard disc to continue your development work, although it is possible that the eventual `.EXE` files will run from floppy.

To save having to set these options each time you run Turbo Pascal, you can save them on disc by using the `Options` sub-menu `Save Options` (or, alternatively, use the program `TINST` to build them into the Turbo Pascal compiler itself).

You are now ready to compile and run a Toolkit program. For your first try we suggest 1-1BISEC.PAS which should already be on your working disc in drive B: (assuming you followed the instructions earlier). This is a good test to start with because it does not use any units. To load it and run it, select Turbo Pascal's **File** menu, and press L̄ to invoke the **Load** sub-option. Now type in the file name 1-1BISEC.PAS, and use the **Run** option (A̅l̅t̅/R̅ is the quickest way of doing this).

Then try 1-3ROOTS.PAS, which does use units. After that, choose other examples as you wish, copy them to the working disc, and run them.

If you find the space restricted on your working disc, there is a variant of the above procedure which relies on the fact that once Turbo Pascal has been loaded, there is no need to keep the system/Turbo disc present in a drive (though it will need replacing if you use Turbo's **Help** options). You will, however, need some of Turbo's .TPU files. So you could create a units disc, containing all the .TPU files from Turbo Pascal and the Toolkit that you need, and keep your working disc for source files and the .BGI files. The units disc could use A: instead of the system/Turbo disc.

) Introduction

Before using the *Scientific Programmer's Toolkit* software we advise you to read this manual thoroughly from cover to cover.

No? Well, seriously, not only do we expect you to ignore the above advice, but we would even like to encourage you to try out the software as soon as possible. But we do strongly advise you to make copies of your *Scientific Programmer's Toolkit* discs, and use the copies as working master copies. Follow the *Guide to installing the software* which is at the front of the book, just after *Acknowledgments*. Then try out some of the Toolkit software, starting with some of the 'ready-to-run' executable programs, and going on to compile and run some example source-code programs (see *Guide to running the software*). In suggesting that you try out Toolkit software sooner rather than later, we are assuming that you have a little experience with DOS and Turbo Pascal, but you do not need to be an 'expert'.

Our purpose in encouraging you to browse through the software a little is to give you an overall impression of the *Scientific Programmer's Toolkit* and the kinds of program that can be written using it. This is also the purpose of the next section, to which we suggest you return after seeing some of the software. Other sections in this Introduction include *How to use this manual*, *How to use the examples*, and *Further reading*, which suggests sources of information on DOS, Turbo Pascal, and Numerical Analysis.

) 1 Overview of the Toolkit

The *Scientific Programmer's Toolkit* has been written for use with Turbo Pascal, the *de facto* standard Pascal system for IBM-PC and compatible machines. Although Turbo Pascal is not fully compatible with the official international standard ISO Pascal, the Turbo Pascal system excels in user-friendliness, combining as it does an editor, compiler, link-loader, screen and graphics drivers, and run-time error system, and it offers such advantages that most users are prepared to put up with the occasional inconvenience caused by lack of ISO compatibility. In writing the *Scientific Programmer's Toolkit* we have taken into account the need to write portable programs which will run under other Pascal systems, and Appendices 1 and 2 deal with this aspect more fully.

The *Scientific Programmer's Toolkit* will be useful to anyone writing programs relating to any quantitative science or technology. At first sight the obvious description of the *Scientific Programmer's Toolkit* is that it is **a collection of mathematical and graphical routines together with utility and user-interface routines**, but in fact the Toolkit offers far more than that: it provides a **framework for writing programs.**

The *Scientific Programmer's Toolkit* **is an aid to software engineering.** The Toolkit gives you both an approach to the overall design of a program or suite of programs, and many of the software components needed. Because the Toolkit is a framework, it is also suitable for use with more specialist graphics or numerical libraries.

The programs and routines provided by the *Scientific Programmer's Toolkit* fall into three categories: *screen text and graphics; mathematical routines; utilities and user-interface.* Cutting across the categories is the concept of *Levels.* The highest level is Level 3 and it consists of complete, ready-to-run, stand alone programs; these have already been compiled and can therefore be used without any compiler, and require no programming effort by the user. You may already have tried out some of these, and they can all be run from the **MENU** program on the Level 3 Toolkit disc (see *Guide to running the software*). Level 3 serves two purposes. The first is to act as a mathematical and graphical calculator for a wide variety of common mathematical tasks: for example, to plot a function or surface, find roots, solve differential equations, draw a smooth curve through data, invert a matrix, *etc., etc.* The second purpose is to demonstrate all the features of the Toolkit.

Level 2 routines are 'off-the-peg' routines that can be incorporated into your own programs with a minimum of fuss. For example, you might be writing a program which involves a certain function, and you might wish to see how the function behaves by plotting its graph. A single Toolkit routine will do this for you, and relieve you of the task of working out scale factors, axis ranges, *etc.* There's no such thing as a free lunch though, and if you use such a routine you have simplicity but you have paid for it by surrendering control over the details of the graph. Therefore we also give you the choice of a more hardworking lunch: Level 1, which gets you to do some of the cooking although many of the ingredients are already prepared for you. Level 1 gives you more control by breaking down a task like function plotting (for example) into a number of component tasks. There may be several parameters associated with each task, and these are now under your control and so you can achieve different effects from those of Level 2. You could even prepare your own Level 2 routines in this way, and store them in your own pre-compiled unit library for later use.

The level structure makes the Toolkit very simple to use, and also provides a learning path by which you can gradually familiarise yourself with the *Scientific Programmer's Toolkit* and its facilities.

Underpinning all the higher levels is Level 0. We often refer to the Level 0 routines as *primitives*. They are an interface to the machine operating system and graphics drivers. For example, although in the Turbo Pascal implementation we have used many of the facilities supported by the Turbo Pascal Graph unit (such as a routine to draw a line between two points specified in screen coordinates), in the Toolkit code calls to such routines occur *only* in the two Level 0 Toolkit units Cit_core and Cit_prim. In higher levels we only use the Toolkit routine Gp_line. This means that if you wish to implement your code on some other computer or Pascal system, it is just a matter of changing the primitives: we have tried to ensure the primitives only do things that almost any computer can do, so that re-writing is almost always possible. Appendices 1 and 2 have all the details.

) 2   How to use this manual

You will already know from the preamble to this *Introduction* that you do not have to read the whole manual from cover to cover before being able to make use of the Toolkit. In this section we suggest how you might approach using the Toolkit book.

The key to finding your way around is to get an overview of what each chapter does. There are several ways of doing that, and the next section (*How to use the examples*) is one. Another way is to use the files in the \DOC directory, and we say more about this in the *On-line documentation* section below.

Chapter 1 is specifically to help you discover 'what to use to do what'. The Pascal concept of units and how they are used is fully explained, and a summary of each unit and what it does is also given. The units are very closely linked with the chapters, and in Chapter 3 onwards the units are described, each chapter dealing with one or more units. As you gain familiarity, you'll get to know which unit contains the type of facility that you want, and this will lead you to the right chapter.

The chapters themselves are also structured. Each section deals with a group of related routines or facilities, and usually begins with a sub-section giving a general introduction to that group. There will be a detailed specification of how to call each routine, and there will be listings of examples of programs using the routines.

Once you have become familiar with the Toolkit, you will begin to want more concise documentation. There are files in the \DOC directory which provide this, and we say more about them in the *On-line documentation* section below.

⟩ 3   How to use the examples

The example programs are a vital component of the *Scientific Programmer's Toolkit*. Almost every section of the book refers to a relevant example, and every feature of the Toolkit is illustrated in at least one of them. The examples are stored in the **EXAMPLES** directory on the examples disc in sub-directories according to chapter. Since each chapter principally relates to particular units, another way of navigating round the Toolkit is to browse through the examples and then refer back to the chapters.

We have made a particular point of avoiding fragments of program. That means that almost every example is a complete working program, and although it is stored on the disc to save you having to type it yourself, you could do so if you wished. Because of this policy, some of the examples may appear rather long for what they do, but we believe that they are much more useful like this.

We intend that many of the examples should be useful as templates for your own programs. We stress that it is good programming practice to have an initial design for your programs before starting to code them, but then you could select an example whose user-interface most closely resembles what you want to do and adapt it to your own design. For this purpose, the primary topic of the example might not be relevant. For example, **CHAP4\1-3ROOTS** might be a reasonable template for a program requiring expressions and graphs even if it is not intended to find roots.

It may be that you can make hybrids from two or more examples; the Toolkit is rich in cross-connections between the chapters.

The generous provision of examples will also be useful if you are building up experience of programming in Pascal. The *Scientific Programmer's Toolkit* does not of course attempt to teach Pascal programming from scratch, but we do not assume that you are an expert, only that you are familiar with the basic structure of a Pascal program and with the mechanics of running a program. Once you have reached that level, we think that learning through example is one of the most effective ways to learn, and hope that in conjunction with a suitable text-book you will find the examples very useful.

) 4   On-line documentation

Once you are beginning to use the Toolkit and have a general idea of its structure, you will want to be able to look up details (such as the signature of any routine) as quickly as possible without having to search through the book. To help you do this, the directory \DOC contains three documentation files: PUBLIC.TXT, TOOLKIT1.DOC, and TOOLKIT2.DOC.

PUBLIC.TXT is simply a concatenation of all the interface sections from the source code of all the units. It therefore contains a list of all variables and routines which you can use, together with their signatures (argument specifications). Even though this is not a complete Pascal program, you can load it into the Turbo Pascal editor and then scan or search the text using standard editing commands. You could also print out this file and annotate it according to your own preferences.

The files TOOLKIT1.DOC and TOOLKIT2.DOC contain signatures of the Level 1 and Level 2 routines, respectively, plus some additional comments. These are a more concise form of the documentation which is printed in this book. Again, you can either look this up whilst working on a computer, or print the files out for off-line reference.

We give a hint which we have found to be very useful in our own experience of using the Toolkit. Turbo Pascal allows you to cut-and-paste blocks of text between files, using the Ctrl/K W and Ctrl/K R commands. It is often helpful to paste into your program a copy of the signature of a routine which you are about to use. You can enclose the signature between comment characters, (*...*) for example, to remind you what all the arguments are.

For those who have the source code of the units, there is a program called **PUBLIC.PAS** which extracts the interface sections from the source code of the units to build the file **PUBLIC.TXT**.

⟩ 5   Further reading

Your computer may well have been supplied with a good DOS manual. Failing that, one of the very many DOS books now available is:

- *MS-DOS and PC-DOS: User's guide* by Peter Norton (Prentice-Hall, 1984).

If you already know a programming language, and need to know Pascal, we recommend:

- *Quick Pascal* by D.L. Matuszek (Wiley, 1982).
- *Pascal User Manual and Report* by K. Jensen and N. Wirth (Springer-Verlag, 1978). [The classic text on Pascal.]

If you have not yet learnt a programming language, and want to gain 'hands-on' experience with Turbo Pascal:

- *Turbo Pascal Tutor Toolbox* (Borland International, 1987).

For more about the mathematical theory of numerical methods, see:

- *A Simple Introduction to Numerical Analysis* by R.D. Harding and D.A. Quinney (Adam Hilger, 1986).

For more advanced texts on numerical analysis, see for example:

- *Computer Methods for Mathematical Computations* by Forsythe, Malcolm and Moler (Prentice-Hall, 1977).
- *Computational Mathematics - an introduction to numerical approximation* by T.R.F. Nonweiler (Ellis Horwood, 1984).
- *Numerical Analysis* by R.L. Burden, J.D. Faires and A.C. Reynolds (Prindle, Weber and Schmidt - Wadsworth, 1981).

⟩ 6   Problems and suggestions

Although we have taken every care to get it right, in a book and software pack of this size there are bound to be a few mistakes, and also places where we have been less clear than we would like in explaining the Toolkit. We would therefore like to ask any user of the Toolkit who has noticed mistakes or has suggestions for improvements to write to us through our publishers.

⟩ Chapter 1

⟩ Introducing the Toolkit

⟩ 1.1   Preliminary

We have designed the Toolkit to be used by people with widely different amounts of computer knowledge. It includes some ready-to-run programs, which require no programming skill, and solve a number of useful problems. But the Toolkit also provides experienced programmers with many flexible and powerful facilities. Much of the Toolkit lies between these two extremes, with the more powerful facilities requiring more skill to exploit. Regardless of your computer experience, we hope you will still find the Toolkit very useful.

⟩ 1.2   Toolkit levels

To represent the diversity of code within the Toolkit, we have divided it into **levels**, with the higher levels being the simpler to use. The Toolkit is divided into four levels, as follows:

- **Level 3** consists of the ready-to-run programs. You do not need any computer experience to use these, and can run them on any PC-compatible system, regardless of whether or not Pascal is available.

- **Level 2** consists of Pascal units. These contain code to solve mathematical problems; to draw a variety of graphs; and to display 'windows' on the screen. They are designed to be quick and easy to use. For example, the Level 2 procedure **draw_curve2** draws a graph of $f(x)$ against x, given only f, and a few other parameters.

- **Level 1** consists, again, of Pascal units. These contain more basic code, which, although more flexible than the Level 2, requires more effort to use. For example, the Level 1 graphics routines allow you to draw far more general graphs (for example, in non-orthogonal coordinate systems). However, they cannot plot a function; you have to use much more basic facilities which can only draw straight lines.

- **Level 0** consists of the Pascal units used internally by the Toolkit itself; although these contain useful code (such as a **tan** function), they are always used in conjunction with the other levels.

The Toolkit is designed so that code from various levels will work together; it is quite in order, for instance, to perform a calculation with a Level 2 mathematics routine, and plot the results using a Level 1 graphics routine.

Let us now introduce the levels in more detail.

⟩ 1.2.1   Level 3

Although the entire Toolkit is written in the programming language Pascal, you do not need to know Pascal (or, indeed, any other language) to run Level 3 of the Toolkit. Level 3 code is ready to run on any computer compatible with an IBM-PC; it does not even need to have a Pascal system available.

If you only intend to use Level 3 of the Toolkit, you can stop reading this chapter now, and move immediately to Chapter 2, which describes the Level 3 code in detail.

⟩ 1.2.2   Level 2

Level 2 of the Toolkit contains the 'building blocks' from which Level 3 is built. Having access to them enables you to solve problems of your own choosing, and modify the Toolkit's behaviour precisely as you wish. Because Level 2 of the Toolkit is written in Pascal (as are the lower levels), your computer must run the Turbo Pascal system, version 4.0 or above, and you will need to do some programming yourself. But the Level 2 code has been designed to be easy to use, and simple to incorporate into your own code. Level 2 is divided into seven units, which, in turn, relate to three different areas: numerical mathematics; graphics; and windowing. We describe the units in this order below:

- Code for *numerical mathematics* is to be found in the units **Cit_math** and **Cit_matr**, which are fully described in Chapters 4 and 5, respectively.

- **Cit_math** provides a number of routines for scalar mathematics. These find the zeroes and minima of functions; fit splines to curves; evaluate integrals (with several different methods available); solve ordinary differential equations (again, with a choice of method); and calculate fast Fourier transforms.
- **Cit_matr** contains code for matrix mathematics. ·This includes the solution of linear equations; decomposition and inversion of matrices; and the evaluation of matrix eigenvalues and eigenvectors. Once again, a choice of methods is provided for most of these operations.

- The *graphics routines* are to be found in the unit **Cit_draw**, which is fully described in Chapter 3.
  - **Cit_draw** contains routines to plot functions or data values in two and three dimensions; draw contours and surfaces; and sample function values in one, two or three dimensions.

- The code for *windowing* is divided between four units: **Cit_ctrl**, **Cit_disp**, **Cit_edit** and **Cit_menu**. These units provide a convenient set of default windows, and facilities to display, modify, and input values within them. A full description of these units is given in Chapter 6.
  - **Cit_ctrl** contains code for program control in a windowing environment. It declares a convenient set of default windows; displays error messages and help information; and provides the 'option menus' from which the user can make a choice.
  - **Cit_disp** contains routines which display numbers and text within a window, with full control over the format of the output.
  - **Cit_edit** contains code which either inputs or edits values within windows. These values may be numbers, strings, or expressions.
  - **Cit_menu** provides a 'menu' facility, from which the user can select a number of options. The help facilities in **Cit_ctrl** can be called automatically if required.

⟩ *1.2.3   Level 1*

Level 1 of the Toolkit consists, again, of Pascal code. The Level 1 code is just as easy to incorporate into your programs as Level 2 code, and requires no more knowledge of Pascal. The difference is that the *facilities* provided by Level 1 are at a more primitive level than those provided by Level 2. As we said earlier, the Level 2 graphics code contains a procedure which plots a function; but the Level 1 graphics code can only draw straight lines. Obviously, a function-plotting procedure can be written which draws a number of straight lines to approximate the required curve (and this is exactly what the Level 2 routine does); but it requires more effort on your part.

The advantage of using the Level 1 code is that you have far more control over the results, and can produce effects which are not possible using Level 2: you could, for example, plot a function in a non-orthogonal coordinate system. Level 1 consists of five units, which we describe briefly below:

- `Cit_eval` contains code which evaluates an arbitrary expression, held in a string, and returns its value.
- `Cit_form` contains routines which display a floating-point value as a string in a number of styles; they provide a wider selection of formats than the Pascal procedure `Write`.
- `Cit_grap` contains the lower-level graphics routines; these draw lines, axes and polygons, and convert between user and screen graphics coordinates.
- `Cit_text` contains routines for handling strings of arbitrary length; a Pascal **string** variable has a fairly small upper bound to the number of characters it can hold.
- `Cit_wind` contains the lower-level windowing routines; these define, save and reload windows, and provide simple operations for writing and reading values within them.

`Cit_grap` is described fully in Chapter 7, and the rest of Level 1 in Chapters 8 and 9.

If you compare the summaries of Levels 1 and 2 of the Toolkit given above, you will see that Level 1 provides more basic, but far more flexible and powerful, facilities.

⟩ 1.2.4   Level 0

Level 0 of the Toolkit is used internally by the Toolkit, and is not designed for use on its own (although it contains useful code like a **tan** function). Level 0 consists of two units:

- `Cit_core` contains the 'core' code, and defines certain types which are basic to the Toolkit, and a few useful arithmetic functions like **tan** and **arcsin**.

- **Cit_prim** contains the graphics primitives, which control the screen at the lowest level.

Cit_core is described in Appendix 1, and **Cit_prim** in Appendix 2; both units declare so many entities that they deserve separate treatment.

Many types and constants fundamental to the Toolkit are declared in **Cit_core**; it is almost impossible to use the Toolkit without them. Because of this, they are described in Section 1.5 of this chapter, in addition to Appendix 1.

⟩ **1.3   Using the Toolkit from Pascal**

If you intend to use only the Level 3 programs, you need read no further; Chapter 2 describes them in detail. (Remember that the Level 3 programs do not require you, or your computer, to understand Pascal at all!) If, however, you wish to incorporate the Toolkit code into your own Pascal programs, you should read on.

The Toolkit, as far as possible, makes use of good Pascal programming style. This means that it relies heavily on Pascal **type** and **const** declarations, and includes some of the more advanced facilities, like **record** and pointer type declarations. However, a fundamental principle of the Toolkit is that it should be easy for you to use, regardless of how difficult it was for us to write! Accordingly, you do not need to understand either **const** or **type** declarations to be able to use the Toolkit effectively; types and constants are far easier to use than to declare.

The Toolkit, therefore, does not merely use predefined types like **integer** or **real**, but makes extensive use of its own types (for example, all the floating-point values you supply to the Toolkit are of type **Citreal**, which the Toolkit defines). Similarly, the **const** declaration is used throughout to provide named constant values, so that one refers to the colour green as **green**, rather than as **2**. This makes code written using the Toolkit much easier to understand (**green** far more obviously represents the colour green than does **2**). Moreover, it is simple to change these definitions if necessary (for instance, if your computer has a non-standard graphics adapter which displays green for colour number 3, rather than 2).

We have used the principle of *information hiding* extensively in the Toolkit. That is, we hide the internal workings of the Toolkit, as far as possible, from the programmer. This is not done out of malice or secrecy;

you can always read the Toolkit source if you wish, to see how something works. But there are two great advantages to information hiding. The first is that you do not need to concern yourself with the internals of the Toolkit in order to be able to use it. Secondly, you are protected from potentially dangerous access to the Toolkit; if, for instance, your program changed values of the Toolkit's pointers, you could well end up crashing your computer. Information hiding prevents you from doing this, and so makes the Toolkit more robust.

To implement information hiding, and to define all the types and constants we require in the Toolkit, we have made extensive use of Pascal *units*. If you are unfamiliar with units, the next section will introduce them. Units are very straightforward to use, in keeping with our philosophy that the Toolkit should be easy to use, even if it uses advanced Pascal techniques. If you already know about units, then you can skip the next section, and read Section 1.5, which describes the basic Toolkit types.

⟩ 1.4   How to use a unit

The simplest way to explain the use of Pascal units is to give a number of examples. This section assumes that you have installed the Toolkit as described in the *Guide to installing the software*; if you have not, you should follow the instructions given there before proceeding.

Perhaps the simplest example we can give uses the procedure **Pause**, which is defined in unit **Cit_ctrl**. The simplest Pascal program to call this would be:

```
program show_pause;
begin
  Pause
end.
```

Obviously, if we try to compile this program as it stands, we shall get an error message, because **Pause** has not been declared.

We do know, however, that **Pause** is declared in the Level 2 unit **Cit_ctrl**, and we need to tell the compiler to search this unit for the declaration. This is done by means of a 'uses clause':

```
uses Cit_ctrl;
```

which should be placed immediately after the **program** statement. This gives us the following program:

```
program show_pause;
  uses Cit_ctrl;
begin
  Pause
end.
```

When the compiler sees **uses Cit_ctrl**, it will scan the declarations in the Toolkit unit **Cit_ctrl**; since this contains the declaration of **Pause**, the program will compile and run successfully.

If you are new to units, it would be well worthwhile compiling and running the program above, for two reasons: firstly, to convince you that units are easy to use; and, secondly, to become familiar with incorporating them into your programs. Use the Turbo Pascal editor to type in the program above, and run it by using [Alt]/[R] (that is, pressing the [R] key whilst holding down [Alt]). (On later versions of Turbo Pascal, you will need to press [R] after using [Alt]/[R]; refer to your manual if necessary.)

When the **show_pause** program above is run, the screen will clear, and a blue window appear at the bottom of the display, saying:

Press any key to continue

If you now press any key, the program will finish, and you will be returned to Turbo Pascal; some versions may first produce the message:

Press any key to return to Turbo Pascal

Obviously, this is a rather useless program, but it does show that the mere addition of:

uses Cit_ctrl;

has made Toolkit code available to you.

'Uses clauses', like the one above, must appear *immediately* after the **program** statement*.

---

* If you wish to use a uses clause within a unit declaration, the uses clause must immediately follow the reserved word **interface**.

Let us now consider a slightly more complicated example. Pascal, unlike many other computer languages, does not have a facility for performing exponentiation (that is, calculating x^y for general x and y). The Toolkit corrects this deficiency by providing a function **Raise**:

```
Raise (x, y)
```

returns x^y, provided that it is real and finite. **Raise** is defined in the Toolkit unit **Cit_core** (this unit is rarely used on its own, but **Raise** makes a very convenient example, so we shall use it anyway). **Raise**, however, has the following Pascal declaration:

```
function Raise (x, y: Citreal) : Citreal;
```

that is, its parameters are of type **Citreal**, and it returns a result of type **Citreal**. Fortunately, this type is also defined in unit **Cit_core** (**Citreal** is so fundamental to the Toolkit that it is defined at Level 0). A suitable program to use **Raise** would therefore be:

```
program show_raise;
  uses Cit_core;
  var x, y : Citreal;
begin
  Write ('Please input two numbers: ');
  Readln (x, y);
  Writeln (x, ' ^ ', y, ' = ', Raise (x, y))
end.
```

Try typing in this program, and running it as before. In response to the prompt:

```
Please input two numbers:
```

enter two numbers (call them x and y). The program will then print out x, y, and x^y. (A good initial choice of x and y would be 2 and 3, respectively.)

In this example, **uses Cit_core** has introduced two entities into our program from the Toolkit: the type **Citreal**, and the function **Raise**. The uses clause actually includes *all* the declarations from the unit named:

```
uses Cit_core;
```

in fact includes more than ninety declarations: procedures, functions, constants, types and variables. In fact a uses clause can incorporate into your program *anything* (apart from a label) which you could declare yourself. We can now see that another advantage of using units is brevity: it is far easier to write:

```
uses Cit_core;
```

rather than type in ninety-plus declarations yourself!

⟩ 1.4.1   *Using more than one unit*

Consider the following program, which plots a graph of $\tan \theta$ against θ for $0 \le \theta \le 1.5$ radians:

```
program show_graph;
begin
  Graphics_mode;
  Draw_curve1 (@Tan, 0, 1.5, 100, CITYELLOW, AUTOAXES)
end.
```

Once again, this program cannot be compiled and run as it stands, because Graphics_mode, Draw_curve1, Tan and CITYELLOW are not defined by Turbo Pascal. The problem is that they are all defined in different units: Graphics_mode in Cit_prim; Draw_curve1 in Cit_draw; and Tan and CITYELLOW in Cit_core. Each program is only allowed one uses clause, and yet we need to use *three* of the Toolkit units.

Fortunately, there is a simple solution to the problem. The reserved word **uses** in a uses clause may be followed by several unit names, separated by commas; so the appropriate uses clause for the program above is:

```
uses Cit_core, Cit_draw, Cit_prim;
```

The order of the unit names is irrelevant; the clause might just as well read:

```
uses Cit_prim, Cit_core, Cit_draw;
```

with exactly the same effect.

You can check this quite simply by typing in the program above, and running it, after incorporating either of the uses clauses above. The results will be identical.

There is no restriction on the number of units which can appear in a uses clause, and they can appear in any order*.

⟩ 1.4.2   Compiling rather than running

So far, we have dealt with running units immediately after they have been compiled, using ⟨Alt⟩/⟨R⟩. How do we save the compiled code to disc, for running later?

The correct procedure is to set the destination of the compiled code to disc, and use the Turbo Pascal **Make** command. To set the destination correctly, use ⟨Alt⟩/⟨C⟩ to enter the **Compile** menu, and press ⟨D⟩ until **Destination Disk** appears on the menu. You can then compile your program either by pressing the function key ⟨F9⟩, or by selecting the **Make** option from the **Compile** menu (by using ⟨Alt⟩/⟨C⟩, followed by ⟨M⟩).

This will compile the program, and write the executable code to disc; you can then execute this merely by typing its file name.

⟩ 1.5   Fundamental Toolkit definitions

The unit **Cit_core** defines many items which are quite fundamental to the Toolkit. Most of these are used only by one unit, and so we describe them with that unit (for example, the type **Expptr**, used by the evaluator, is documented with the evaluator unit **Cit_eval**).

However, some of the items are used throughout the Toolkit, in several units; we therefore introduce them briefly here. Please note that fuller descriptions are given in Appendix 1 under **Cit_core**, to which you should refer if you need more information.

⟩ 1.5.1   Real types

The Toolkit uses the type **Citreal** wherever it needs a floating-point value; likewise, any parameters passed to the Toolkit which represent

---

* Note that the manual for Turbo Pascal, version 4.0 is wrong here; the correct documentation for units appears in the **README** file for this version.

floating-point values are of type `Citreal`. The Toolkit does this quite deliberately, rather than using one of Turbo Pascal's built-in types (like `real`) directly.

The reason for this apparent cussedness is that the Toolkit can easily be changed to use *any* of the Turbo Pascal types merely by changing the definition of `Citreal`. By default, if you have no numeric coprocessor, the type `Citreal` is identical to Turbo Pascal's `real`; if you have a numeric coprocessor, `Citreal` is identical to Turbo Pascal's `double`.

In addition, the Toolkit defines the constants `MAXREAL` (to the maximum value which can be held by a `Citreal`), `MINREAL` (to the minimum value which can be held by a `Citreal`), and `EPSILON` (to the machine epsilon* for a `Citreal`).

⟩ *1.5.2   Colours*

`Cit_core` defines names for the colours which can appear on the computer screen; these names consist of the colour name, prefixed by `CIT`, as follows:

Colour	Represented by
black	CITBLACK
blue	CITBLUE
green	CITGREEN
cyan	CITCYAN
red	CITRED
magenta	CITMAGENTA
brown	CITBROWN
light grey	CITLIGHTGRAY
dark grey	CITDARKGRAY
light blue	CITLIGHTBLUE
light green	CITLIGHTGREEN
light cyan	CITLIGHTCYAN
light red	CITLIGHTRED
light magenta	CITLIGHTMAGENTA
yellow	CITYELLOW
white	CITWHITE

---

* The smallest value ϵ, such that $1.0 + \epsilon \neq 1.0$, when evaluated by the machine.

Please note that the spelling for the **CIT** greys follows the standard for IBM-PCs, which is not the same as the British standard!

⟩ *1.5.3   Linestrings*

The type **Linestring** is a Turbo Pascal string which can hold a maximum of 80 characters—the maximum length of a line on the screen in text mode.

⟩ *1.5.4   Errors*

When the Toolkit code detects an error, it sets three global variables, defined in **Cit_core**. The Boolean **Errorflag** is set to **true**, to indicate that an error has occurred; the integer **Errorcode** is set to a number which reflects the type of error; and the string **Errorstring** is set to a brief description of the error.

⟩ Chapter 2

⟩ Level 3 Programs

In the Toolkit there are programs and routines. The programs run independently of the Turbo Pascal working environment on any PC-compatible machine with standard screen adapters. They represent sets of Toolkit routines constrained to different degrees to suit predefined circumstances.

In contrast, the routines at Levels 1 and 2 have to be called from a Pascal program written by the user and therefore are to be used from within the Toolkit environment. Both the programs and routines work on the same principles, manipulating and converting standard objects such as real values, arrays, text, expressions, images and windows. It is the method of access that differs.

This chapter describes the four Level 3 programs, TK, LAPLACE, ODE1, and FSCOEFF, and provides a useful introduction to the facilities of the Toolkit as a whole with particular reference to applications within a teaching environment. It is not concerned with the detail of programming, although there is some discussion of the overall design of the code.

⟩ 2.1   The Level 3 programs

The program TK is general purpose, and contains nearly a full set of Toolkit facilities. It is suitable for general computation and might be established as a library facility. It is a large program, and has its own working environment within which the main Level 2 mathematical and graphics routines can be selected and run. Data can be named, using the conventions described for Level 2 in Chapters 3 and 4. The environment itself is constructed from the window, text and control routines found in Levels 1 and 2 and described in Chapters 6, 7 and 8. Overall, this program is a useful and quick tool for calculations, and it is also a shop window for the individual routines, showing how they work and how data might be prepared.

The three programs, LAPLACE, ODE1 and FSCOEFF, have a purpose which is much more closely defined. They might be used while teaching or learning specific subject material and provide illustrations of what can be achieved using the Toolkit. Three situations are considered:

- investigative projects, where students write and run their own programs,
- class exercises, where teachers write the programs for the students' use in the laboratory, and
- class demonstrations, where teachers write and run the programs. Each program is described within its mathematical context.

When the Level 3 programs run you will see that they write and read data at intermediate stages to and from ASCII files. These files have names with the .DAT extension, and contain information on objects created. They are usually referenced when editing objects and contain default expressions and data sets. The data can be prepared beforehand by the user from the Level 1 and 2 routines, or they might be constructed by a word-processor. You should find the formats straightforward; details are given at the end of this chapter.

⟩ 2.2   Running the Level 3 programs

The file TK.EXE, together with the small .BGI files, is needed to run the program TK. These files are to be found on the disc labelled TK.

The file DEMO.EXE, together with the small .BGI files and the .HLP files are needed to run the other three Level 3 programs, LAPLACE, ODE1, and FSCOEFF. They should be copied into a directory DEMO on your hard disc. (If you are not sure which .BGI files are needed, copy them all.) The programs can then be run by entering the command DEMO and making selections from the menu.

It might also prove useful to copy the sets of .DAT files into the current directory. This will be useful the first time the programs are used by providing default values for the parameters.

There is Pascal code for the set of three Level 3 programs on the same disc. The source files have similar names, with extension .PAS. This code can be amended to suit other situations. It will then, of course, be necessary to re-compile it using the Toolkit.

⟩ 2.3   The program TK

The program TK provides direct access to the main mathematical and graphics facilities. It is a shop window for the Toolkit's facilities.

ACTION. Run the program TK, and press any key to move to the working display.

Now look at the options in more detail.

⟩ 2.3.1   *Windows and the arrangement of the screen*

By convention, there is an *Instruction* window along the bottom of the screen, and a *Menu Bar* along the top of the screen, leaving the middle for *Display*. Other windows will appear and disappear over the display area. In particular, *Application* windows will be opened and closed in the display area. Several of these can be open at the same time, each one containing an application of one or more of the Toolkit routines. Each application window carries its own set of data values associated with the parameters that are specified for the routines.

As new applications are opened the windows will be arranged on the screen in one of four patterns, *cascade*, *tile*, *full* and *user*:

• In the cascade pattern, windows are overlaid in a line from the top left of the display area. (This is what will happen by default.)
• In the tile pattern, windows are set side by side, each window taking a quarter of the display area. If more than four windows are open they are overlaid.
• In the full screen pattern, the window for each application fills the display area, completely covering what was there before.
• In the user pattern, the window size and position are determined by the user.

There is a window management system built into the program TK and this gives the user control over saving and recalling images contained in the windows.

ACTION. If you want to change the pattern in which the windows are arranged, choose the option **Screen** on the main menu (use the left cursor key and press $\boxed{\texttt{Enter}}$, or just press $\boxed{\texttt{S}}$) and then select a pattern (use the down cursor key and press $\boxed{\texttt{Enter}}$, or just press the first letter of a menu option).

When new windows are opened the screen is partially saved in memory, to be restored when the new window is deleted or moved to the background. Thus, the amount of memory consumed by the patterns varies. This has to be an important consideration, particularly for the higher-resolution graphics modes. With the VGA screen, for example, there will not be enough memory in the 640K DOS area for more than two application windows in the 'full' pattern, or more than five windows in the 'cascade' pattern.

It is a feature of the Toolkit that memory is used to store window images when possible up to a limit. The limit is set to ensure that there is some memory always available for numerical and housekeeping tasks. Thereafter the images are saved to disc. This takes longer but, if a hard disc is available, there will be far more scope for keeping many windows active. Should there not be enough space on the disc then a warning message will be displayed.

ACTION. If, at any time, you want to list, close, or open windows, choose the option **Window** on the main menu, and then select an option from the window menu.

You can save windows to disc files and load them again later by choosing from the **Window** menu. The files created have the default extension .WND.

⟩ *2.3.2   Tools*

The Toolkit facilities are called *tools*.

ACTION. Select the option **Tools** in the main menu.

You will see a list of options divided into three sections: *Graphics, Numerical* and *Matrix*. These are the three main categories of the Toolkit computational facilities, and the options relate to the Level 2 routines found in Chapters 3, 4 and 5, respectively.

ACTION. For example, select the **Plot curve** option (use the up/down cursor keys and press Enter).

When a curve is plotted, data are required in the form of mathematical expressions and axis variables. There are three combinations available for formats, and these are shown in the sub-menu. The first specifies a curve explicitly, in terms of $y = f(x)$, y being plotted against x. The second format is parametric and operates in two dimensions, $y(s)$ being plotted against $x(s)$ for a range of values of the variable s. The third is parametric in three dimensions, points $(x(s), y(s), z(s))$ being plotted over a range of values of the variable s.

ACTION. For example, select y=f(x).

An 'application' will now have started. As, until now, there was no application window active, a new one is opened and displays a graph on default axes. In addition, a parameter window is opened. In this window the parameters for the associated routine are listed. For the above selection, the routine is **Draw_curve**. Whatever is now specified will be attached to this application of the routine **Draw_curve**.

The names f and x used above are only notional, and provide a quick description of the objects needed for the routine. These names are given initially by default, but they can be changed. If they are, the new names will become the defaults for the application. You may want to change them to correspond with an object already in memory. It could be an expression, say, named $g(s)$, in which case the name g will be adopted instead of f, and the name s instead of x.

ACTION. For the present, press $\boxed{\text{Enter}}$ twice and accept the names f and x.

You now have to specify an expression $f(x)$. For this parameter window only one-line expressions can be typed, though multi-line expressions can be prepared separately. As it is, the line can be up to 255 characters long. The edit window scrolls sideways to accommodate long strings.

The expression taken by default will be the one last used under the name f. If there is not one in memory, then a data file is sought with a corresponding name. (In this case the name will be **F.DAT**; there is a pre-prepared file on the distribution disc of this name.)

The expression can contain functions (as listed for the Level 1 evaluation routines described in Chapter 9) and variables other than those named in the function. These variables will for the time being be treated as constants, and must have been assigned a value in the dictionary. Usually this will be done separately, the variable being assigned a real value under the option **Objects**.

ACTION. Press $\boxed{\text{Enter}}$ to accept the default expression.

The x-range has to be specified. The values entered can take the form of an expression containing constants (e.g. π) and variables which have been assigned a real-valued object.

ACTION. Press $\boxed{\text{Enter}}$ twice to accept the defaults.

Now the draw options are selected. The curve can be superimposed on the current graph, scales can be set automatically, or the y-range can be specified explicitly.

ACTION. Press $\boxed{\text{N}}$ in reply to the question "Superimpose?", and press $\boxed{\text{Y}}$ when asked "Autorange?".

The parameter window disappears, the axes are rescaled and the curve drawn. Then a *Sequel* window is opened. You can now choose to **Repeat** the operation, re-running the routine with a new set of parameter values; to overlay grids and cross-wires to evaluate points on the graph; to remove the sequel window temporarily if it is obscuring the graph; to view the parameter values; or simply to return to the menu bar.

ACTION. Select the **Menu Bar** to return to the main selection menu.

All the tools operate in this way. A sequence of sub-menus, parameter windows and sequel windows take you through an application. Several tools can be used in any application, and a record is kept of the last values used in the application for the parameters. Values of the parameters are also carried forward as defaults when a new application is opened.

A list of tools and the associated options, taken from routines in the units `Cit_draw`, `Cit_math` and `Cit_matr`, follows:

Tool	Options	Routine
Draw curve	`y=f(x)`	`Draw_curve1`
	`x=x(s),y=y(s)`	`Draw_curve2`
	`x=x(s),y=y(s),z=z(s)`	`Draw_curve3`
Plot data	`y[i],x[i]`	`Draw_data1`
	`z[i,j],x[i],y[j]`	`Draw_data2`
	`x[i],y[i],z[i]`	`Draw_data3`
Draw histogram	`y[i],x[i]`	`Draw_histogram1`
	`z[i,j],x[i],y[j]`	`Draw_histogram2`
Draw contours	`z[i,j]` 2dimnl	`Draw_contours`
	`z[i,j]` 3dimnl	`Draw_contours`
Draw surface	`z[i,j]` Wireframe	`Draw_surface`
	`z[i,j]` Surface	`Draw_surface`
Body of revolution	`x[i],z[i]` Wireframe	`Draw_revolution`
	`x[i],z[i]` Surface	`Draw_revolution`
Zeroes of function		`Zeroin`
Minimum of function		`Fmin`
Cubic spline		`Spline,Seval`

Tool	Options	Routine
Integration	Romberg	`Romb`
	Adaptive quadrature	`Quanc8`
Solution ODEs	4th-order Runge-Kutta	`Rk4`
	Adaptive RK-Fehlberg 1	`Rkf`
	Adaptive RK-Fehlberg 2	`Rkf45`
Fourier transform		`Fft`
Linear equations	Gaussian elimination	`Elim`
Inversion	Gaussian elimination	`Elim`
	Decomposition	`SVD`
Eigenvalues	Power	`Power`
	Modified power	`Modpower`
	Jacobi's method	`Jacobi`
Sing. value decomp.		`SVD`

⟩ 2.3.3   Objects

The primary working objects at Level 3 are *real values*, *arrays* and *expressions*. In Level 1, there are other objects named, like parse trees, text and strings, but here these have a secondary role. In fact, here the 'real values', 'arrays' and 'expressions' can all take a number of different forms. As they appear to the user they will be text objects and can be saved to disc, read from disc, and edited. However, when they are used in the applications, the appropriate Level 1 objects are substituted so that they can be quickly accessed in calculations. But this occurs automatically, and the housekeeping associated with the other Level 1 objects need not concern us at the moment.

ACTION. Select the option `Objects`.

In the `Objects` menu there are options to list, edit and delete objects. During editing, the objects can be converted from one type to another in the following ways:

- An expression can be sampled to give a real value.
- An expression can be sampled to give an array.
- An array can be redimensioned.

To serve as an illustration, suppose we wish to sample an expression in two variables $z(x, y)$.

ACTION. Select **Edit NEW OBJECT** and choose to **Sample expression.** An array **z[i,j]** can be sampled over a 15 × 15 grid of points with coordinates $(x[i], y[j])$ with values based on the function **z[x,y] = cos(x^2+y^2)** where **x** is taken from the range **[-2,2]** and **y** is taken from the range **[-2,2]**.

A data array is constructed in memory. It can be inspected using the **Display** option. Then, **Tools** can be selected and a **Surface** drawn based on this array.

⟩ *2.3.4 Params*

The parameters used in the the most recent application can be recalled from the option **Params**.

ACTION. Select the option **Params**.

⟩ *2.3.5 Files*

The files on disc, and the associated objects, can be listed, viewed, deleted and edited in text form.

ACTION. Select the option **Files**.

⟩ *2.3.6 Help*

Help is available at Level 3. Normally, help can be activated by pressing the F1 key whenever there is a message **F1 Help** in the bottom left of the screen. If there are any related help messages, they are displayed. Messages are provided to help you work through other options in the above manner.

ACTION. Select the option **Help**.

Some help messages give advice on which menu option to select. They are called when 'Auto-help' is active. This is not selected by default, but has to be switched on by selecting **Auto on** in the menu. Later it can be switched off by selecting **Auto off**. These auto-help messages persist on the screen and can be a nuisance to someone familiar with the conventions.

Help messages can always be called when selecting menu options. Pressing the F1 key displays a message specific to the option currently highlighted in the menu window. The message will disappear when a cursor key, or Enter , is pressed.

Once called, help messages are retained in memory, where they take up valuable space. However, they can be cleared by selecting the option **Wipe**. If the messages are called later, they will automatically be reloaded from disc.

⟩ 2.3.7   Design of the code

Although the intention of this chapter is to give an overview, rather than detail, of the Toolkit, it is useful to see how, in general terms, the program **TK** has been designed.

The code is included on the distribution disc, and has been arranged to call the Level 2 routines in a way which illustrates their features. Inasmuch as the Level 2 routines all take a set of parameters, operate on objects and work inside a window display, the program **TK** has been split into three main units, one assembling the parameters for the tools (**Tk_tool**), one manipulating objects (**Tk_objec**), and the other manipulating windows (**Tk_wind**). There is also a fourth unit, **Tk_selec**, which presents the program's menus and options.

At the centre is a set of Pascal record types which describe objects and window applications. Each application has associated text which contains the parameter values. When a tool is called from the menu, reference is made to a template which describes the parameters needed. Where possible these parameters are matched to the parameter values used in previous applications. The user is then given the defaults and asked to confirm them or supply new parameter values. The parameter text is updated, and the routine called and supplied with the most recent set of values.

In this way the program's facilities are 'packaged' with operations that correspond to the decisions that the user must make when using the Level 1 and 2 routines. The program uses its templates in an analogous way to the user looking up the reference sections contained in Chapters 3, 4 and 5.

⟩ 2.4   The program LAPLACE

The program **LAPLACE** provides an example that might arise in an investigative project. It is a specimen solution written to answer specific questions. It is straightforward in design and operates through simple 'button pressing', the data being embedded within the program code. The topic is the Dirichlet problem, a discussion of which is common in undergraduate work.

The Dirichlet problem is the solution of Laplace's equation over a region of given shape. It is often tackled through Fourier series or through complex analysis. However, shortage of time in mathematics syllabuses and an increasing emphasis on applications make these approaches less appealing, and the tendency now for students of applied mathematics is to place more emphasis on numerical methods. Moreover, the numerical approaches provide convenient stepping stones for engineering students to the Computer Aided Design packages they are now meeting in their engineering coursework.

The aim is to give students a 'feel' for the process of solving the equation and an appreciation of the limitations of the approach. It is fortunate that amongst the numerical methods for tackling this problem there are some that are simple enough to be understood and programmed by most students. For this reason the topic is well suited for use in investigative classwork, where students have to study an algorithm, write a program to implement it, and report their findings.

) 2.4.1   The exercise

The exercise is to write a program which draws a contour map of steady-state temperature over a steel plate in three situations:

1. A square plate with temperature 100°C on one side and 0°C on the other three sides. (This is a classic situation examined analytically using Fourier analysis in O'Neil *Advanced Engineering Mathematics* Wadsworth, Section 13.3.)
2. A circular disc with unit radius and temperature $U(\theta)$ along the boundary of

$$U(\theta) = \begin{cases} 100 & \text{when } 0 \leq \theta < \pi \\ 0 & \text{when } \pi \leq \theta < 2\pi \end{cases}$$

(This is a classic situation examined using complex analysis in O'Neil, Chapter 19).
3. A rectangular plate of dimension 8 × 4 with temperature 0°C along its outer edges. There are two circular holes of unit radius centred on points distance 2 along the longer axis of the plate. The temperature along the boundary of the holes is 100°C.

In the method of finite elements a grid is imposed over the region. We consider it an $m \times n$ grid of size h containing N grid points, where $N > 600$. In the program an array u_{ij} is to be initialised to reflect the boundary conditions. Values at internal grid points are set to zero and those at points on or over the edge are set to the value of temperature along the boundary. The program scans through the array revising the interior elements one-by-one according to the formula

$$u_{ij} = 1/4\left(u_{i+1,j} + u_{i-1,j} + u_{i,j+1} + u_{i,j-1}\right)$$

If and when the values of the array settle, a contour map is drawn.

ACTION. Run the program **LAPLACE**. Select options and obtain specimen solutions for the three parts of the exercise.

All three parts require more than 200 iterations before the solution begins to settle. Coarser grids tend to give faster convergence, but provide a less-precise solution. The program might be modified to change from coarse to fine grids as the solution is approached.

The third exercise has a finer grid, and it takes a long time to work through the 200 iterations. But the region is more complex and it becomes necessary to retain a fine grid to ensure the region is accurately represented. This part of the exercise pushes the humble PC towards its limits.

It becomes natural to question the accuracy of the methods, and ask how the results compare with those that might be obtained analytically. There is an extension to the code already written into the program **LAPLACE**. A Fourier series solution is constructed and plotted over the top of the last contour plot obtained using the Gauss–Seidal method. When you press [Esc] the Fourier series solution is computed automatically. Within the accuracy of the grid, the results are quite close.

A similar facility is available for the second situation. This time the Poisson integral is computed for each grid position, and this solution plotted over the Gauss–Seidal solution. Unfortunately, given the higher level of accuracy required in the numerical integration, and given the relatively large number of grid points, this computation takes a long time. If you get tired, press [Esc].

) 2.4.2   *Design of the code*

The program in file **LAPLACE** is a specimen answer. It is designed to work through the three exercises giving a visual picture of the contour pattern. The program relies on using the program control routines and the graphics necessary to draw the contour pattern. The program flow is straightforward, the data being embedded in the code.

Note that by choosing to draw nine contours in the range 0–100°C, the curves will be placed at exactly 10°C intervals.

⟩ 2.5   The program ODE1

The program ODE1 provides an example of a class exercise in which students are given a worksheet to work through at their own pace in the computer laboratory.

Consider that a lecture course has started with the discussion of First-order Differential Equations. In classical terms this means identifying a number of special forms and working through solutions of equations. It should also lead to an understanding of the concepts of families of solutions, isoclines, and direction fields, and some skill in sketching solutions, particularly of equations that do not have the classical forms.

Student difficulties often lie in two main areas. The first is in being able to visualise isoclines and direction fields, and the second is in learning to sketch. Individuals seem to work at different paces, and everyone needs experience with informed feed-back. In both respects the subject area is well suited to computer-based classwork in the form of worksheets of structured exercises supported by a prepared program, one that has been written in advance by the lecturer.

⟩ 2.5.1   The exercise

The exercise is arranged as a worksheet, to be completed with the help of the program ODE1. The program contains basic facilities and it is for the student to decide how it can help.

WORKSHEET
Solution families for first-order differential equations

1.   The objective is to solve the equation

$$\frac{dy}{dx} = -\frac{x + 2t}{3x^2 + t}$$

over the region $x \in [-1, 4]$, $y \in [-2, 2]$. Arrange the equation in exact form and thereby sketch the family of solutions.

2. (Adapted from O'Neil, page 77, Question 56.) An oil tanker of mass M is sailing in a straight line. At time 0 it shuts off its engines and coasts. Assume that the water tends to slow the tanker with a force proportional to v^α, where v is the velocity at time t.

Resolving forces leads to the following differential equation for v as a function of time:

$$Mv' = -kv^\alpha; \quad M, k > 0, \quad v > 0$$

a. Use an appropriate substitution to solve this equation and thereby find an expression for v in terms of t. Show that the tanker eventually comes to a full stop if $0 < \alpha < 1$, and find the time taken. What happens if $\alpha \geq 1$?

b. Sketch out the family of solutions for the cases

$$\alpha = 0.2, 0.5, 1.0 \quad \text{and } 1.2$$

and for the first two of these derive estimates of the time taken to stop (in units of M/ks) for an initial velocity of 40 m/s. (Of course at the same time you can measure times for all start velocities in the range 0–40 m/s.)

Describe what happens to the velocity in the other two cases.

(Change the notation so that you can use the program ODE1. Select the options to answer the questions. In your answers state what the curves represent. Remember, an overall map of the solution families can always be obtained, if necessary, by using the option Slopes.)

Write the equation in exact form, and thereby derive a function $G(t, v)$ for which the family of solutions are the contours of $z = G(t, v)$.

c. Re-write the original equation in terms of v and s, the distance travelled. Solve this equation and find an expression for s_0, the distance required to stop the tanker for the cases $\alpha = 0.2$ and $\alpha = 0.5$.

A typical laboratory session should include the following work. The results will be obtained with conventional analysis and computer work.

1. The equation can be rearranged in exact form with

$$F(x, y) \equiv y^3 + xy + x^2 = K$$

ACTION. Run the program. Denote the right-hand side (RHS) of the Equation `f(x,y)=-(y+2*x)/(3*y^2+x)` and enter the Exact_sol `F(x,y)=y^3+x*y+x^2`.

Note that the complete family of solutions is displayed both by using the option **Exact_sol**, which draws contours of the function $F(x, y)$, or **Slopes**, which draws short lines at slopes equal to $f(x, y)$. Any particular solution can be plotted using option **Solution**, at least to the extent that a direct application of the Runge-Kutta method can draw the curve. More information about the value of the slope can be obtained using the option **Values**.

There is a region over which the solution $y(x)$ has multiple real values. A number of contours turn back on themselves. They do so because the denominator of the expression $f(x, y)$ becomes zero along the parabola $3y^2 + x = 0$. Nevertheless, particular solutions exist, and can be traced, for all other points in the region.

2. The substitution

$$u = v^{(1-\alpha)}$$

can be used. You will find that when $0 < \alpha < 1$ the solution of the differential equation is

$$v(t) = \left(v_0^{1-\alpha} - \frac{k}{M}(1 - \alpha)t\right)^{\frac{1}{1-\alpha}}$$

and the time taken to stop is

$$t = \frac{Mv_0^{1-\alpha}}{k(1 - \alpha)}$$

When $\alpha = 1$,

$$v(t) = v_0 e^{-\frac{kt}{M}} > 0 \qquad \text{for all } t$$

and if $\alpha > 1$ then

$$v(t) = \frac{1}{\left(\frac{1}{v_0^{\alpha-1}} + \frac{k}{M}(\alpha - 1)t\right)^{\frac{1}{\alpha}} - 1}$$

and so $v(t)$ is never zero.

ACTION. Use the name **y** for the variable v, and **x** for the variable t. Then, in the different cases the expressions $-y^{0.2}$, $-y^{0.5}$, $-y$ and $-y^{1.2}$ can be entered in turn. When prompted, use the names g1(x,y), g2(x,y), g3(x,y) and g4(x,y), and the expressions will appear by default. Use an x-range $[0, 50]$ and a y-range $[1, 50]$. Trace particular solutions by starting at points **y** = 10, 20, 30, 40 and 50, and record the x-values whenever $y = 0$. Use these values to plot a graph (on graph paper).

When $\alpha < 1$ the solutions can be interpreted as contours to the expression

$$z = G(t, v) \equiv v^{(1-\alpha)} + \frac{k}{M}(1 - \alpha)t$$

ACTION. The family of solutions can also be obtained by entering, for example when $\alpha=0.2$, G1(x,y)=y^5+0.8*x.

Note that contours of this expression can only be drawn for values of y strictly greater than zero.

For the other values of α, the corresponding expressions are available by default under the names G2(x,y), G3(x,y) and G4(x,y).

Now re-write the equation using the substitution $v' = v\frac{dv}{ds}$, and find that the distance taken to stop is

$$s_0 = \frac{M}{k} \frac{v^{(1-\alpha)}}{1 - \alpha}$$

Although not necessary in this case it is worth noting that values for distance s_0 can be obtained (in units of $\frac{M}{k}$) by solving a new equation.

ACTION. Enter a new equation. Denote the variable v by the letter **y** and the variable s by the letter **x**. The appropriate **RHS** expressions are available by default under the names h1(x,y) and h2(x,y). Use the option **Solution** to obtain values for s_0.

〉 2.5.2   *Design of the code*

Programs used as tools must have more flexibility. They will have menus which ideally have adequate explanations on screen. It is quite likely that the programs will be used without supervision, and, instead of providing specimen input of the type that helped the lecturer follow a prescribed path, it is better to add help, either directly on the screen, or through a help facility. You will see in the program that help is displayed by pressing F1 when selecting from the bar menus.

The program ODE1 works on a central Level 1 graphics window and operates in two dimensions. It uses Level 2 of the Toolkit to draw contours (routine **Draw_surface**) and draw solution curves (routine **Rk4**). Menus and help are available through routines in the units **Cit_menu** and **Cit_ctrl**.

The program is formed in one controlling loop containing different options. Within each option data have to be supplied before proceeding.

Of note is the use of a **log** transformation when drawing contours. This has the effect of reducing extreme values which will occur, for example, when plotting solution curves near the parabola $3y^2+x = 0$. As a result, the contours will not be equally spaced, but will be concentrated on values $F(x,y)$ near zero.

) 2.6   The program FSCOEFF

The program FSCOEFF provides an example of a class demonstration.

Suppose the subject of a lecture is the ability of Fourier series to reproduce values taken from a given periodic function. At the outset, there is a point to be made about the convergence of the series to an approximated function. Over a wide range of cases, increasing the number of harmonics in a Fourier series will in some sense improve the approximation. In a lecture, this needs to be illustrated and discussed. It is informative to see the nature of the approximation, and appreciate what happens to the error as the number of harmonics is increased.

The difficulty with this material becomes apparent when working through an example. To evaluate the Fourier coefficients, you normally need to work out the somewhat tedious definite integrals that arise in the Euler formulae. Further, the coefficients obtained are, in themselves, obscure, and only become meaningful after the series is evaluated. Essentially, the material calls for a comparison of shapes rather than analytic expressions.

The advantage of a computer-based approach is that the shapes can be reproduced quickly and compared visually. Moreover, the power of the PC is sufficient for the process to be treated dynamically, and this can help spin out a storyline. Thus the computer, and indeed the

Toolkit, is capable of supporting numerical operations which help illustrate and motivate the analysis.

) 2.6.1   *The exercise*

O'Neil provides three worked examples on pages 718–20 (Second edition).

a. The step function

$$f(x) = \begin{cases} -1, & -1 \leq x \leq 0 \\ 1, & 0 \leq x \leq 1 \end{cases}$$

b. The tooth function

$$f(x) = |x|, \quad -\pi \leq x \leq \pi$$

c. The sawtooth function

$$f(x) = x, \quad -\pi < x \leq \pi$$

We shall give these functions, and their sections, the names **f11**, **f12**, **f2** and **f3**, respectively, and investigate the approximations provided by the Fourier series for partial sums:

$$S_m(x) = \frac{a_0}{2} + \sum_{n=1}^{m} a_n \cos \frac{n\pi x}{L} + b_n \sin \frac{n\pi x}{L}$$

where $m = 1, 2, 3, 6$ and 10.

ACTION. Run the program **FSCOEFF** and enter the expression for example **a**. In the first section $f1(x) = $ **f11** over the interval $[-1,0]$. In the second section $f2(x) = $ **f12** over the interval $[0,1]$. First, set $m = 1$. Then select the option **Harmonics**, and superimpose the curves for $m = 6$ and $m = 10$.

The Fourier series can be constructed analytically:

$$f(x) = \sum_{r=0}^{\infty} \frac{4}{(2r+1)\pi} \sin((2r+1)\pi x)$$

The values displayed in the program are the coefficients of this series. The function is 'odd' and there are no cosine terms in the series. Convergence is fairly rapid.

The function has a discontinuity at $x = 0$. Now investigate Gibbs' phenomenon.

ACTION. Use the option In_zoom and change the axis ranges, and focus on the region $x \in [0, 0.5]$ and $y \in [0.8, 1.2]$. Recompute the Fourier series with $m = 20$ and $m = 40$. In each case measure off the screen the largest value of the discrepancy $|S_m(x) - f(x)|$ and the position where it is attained.

Note that the overshoot remains after the discontinuity, even though elsewhere the series moves closer to the function shape $f(x)$.

ACTION. Select the option **Period** and repeat the exercise for example b.

For the second example the function is 'even' and only has cosine terms. The Fourier series is

$$f(x) = \frac{\pi}{2} - \sum_{r=0}^{\infty} \frac{4}{(2r+1)^2 \pi} \cos((2r+1)x)$$

Convergence for this example is quick. There are no discontinuities in $f(x)$ itself, but $f(x)$ is not differentiable when $x = 0$.

ACTION. Choose **Ranges** and investigate what happens in the region $x \in [-1, 1]$, $y \in [0, 0.5]$.

Compare this example with the last and answer the question 'Does the slope overshoot?'.

ACTION. Select the option **Function** and repeat the exercise for example c. Try $m = 3$, **6** and **10**.

For the third example the function is 'odd' and only has sine terms. The Fourier series is

$$f(x) = \sum_{r=1}^{\infty} \frac{2(-1)^{n+1}}{n} \sin(nx)$$

Convergence is much slower.

ACTION. Finally, try $m = 30$, **40** and **50** for example c.

This computation takes a long time and you may have to leave it to cook. The result is a degree of convergence equivalent to that achieved for values of m less than 10 in the other two examples. You can see this by noting the values of the coefficients.

⟩ *2.6.2   Design of the code*

For a class demonstration there is need for clarity in the diagrams (so that students can see what is happening), and simplicity in the input of data (to avoid mistakes on the part of the lecturer). There will not be a need for explanations on the screen, for these come in the lecturer's introduction, and some set examples should be available.

In the program **FSCOEFF** there is one main display area, possibly overlaid by a table of numerical values. The expressions are called by default from .**TXT** files. The calculations progress in a sequence parallel to the theory.

When a function $f(x)$ is entered, it is specified in sections $f1(x)$, $f2(x)$,...and so on. The graphs of $f(x)$ and $S_m(x)$ are drawn using the Level 2 routine **Draw_curve**, the two expressions being evaluated as functions of x. The Euler formulae are evaluated numerically over each section using Romberg integration (Level 2 routine **Romb**), and the value of $S_m(x)$ by summing terms in the Fourier series.

Some tuning of this program is necessary to ensure that the correct balance is achieved between speed and accuracy when computing the values of the coefficients. As arranged, there are default settings for the number of steps in the Romberg integration, and these can be adjusted to give, for example, greater accuracy. However, it should be realised that there is a limit to the precision obtained on a PC in normal arithmetic modes, and this will inevitably limit investigations at some stage.

Within the code the overall set of calculations are embedded in loops. Thus the natural order of the program, after an example has been completed, or a panic Esc has been pressed, is to start again at progressively earlier stages in the calculations.

⟩ **2.7   Preparing data for Level 3 programs**

Data files have the extension .**DAT** and the Toolkit objects are saved and loaded by routines in the unit **Cit_text**. The files are all ASCII text files. The first line contains a description, and subsequent lines contain data.

1. Value.	name	value
	e.g. **A**	**3.21**

2. Matrix.   *name (dimensions)*   values in matrix form

e.g. `A(2,4)`   `11 12 13 14`

`21 22 23 24`

3. Expression.   *name (variables)*   multi-line expression

e.g. `f(x)`   `exp(1+x^2)`

Only when the file name matches and the first line corresponds to the required type are the data read from the file.

Files can be prepared separately using a text-processor or a word-processor, or they can be output from a user program which itself might be constructed using routines from Levels 1 and 2. Thus the default values and expressions used by the programs **FSCOEFF**, **ODE1** and **LAPLACE** can be pre-prepared.

⟩ Chapter 3

⟩ Graphics Routines

⟩ 3.1   Introduction to Level 2

This is the first chapter in the section of the *Scientific Programmer's Toolkit* which deals with Level 2 routines, and so here we recapitulate their purpose briefly: Level 2 routines are 'off-the-peg' routines which can be incorporated into your own programs with the minimum of fuss. There are Level 2 routines for the full range of Toolkit topics: graphics in Chapter 3 (this chapter), mathematical routines in Chapters 4 and 5, and screen and option control using windows and menus in Chapter 6.

Each chapter consists of self-contained sections dealing with a routine or group of routines with a closely related purpose. Each chapter includes an introduction to the chapter topic, descriptions of demonstration programs, detailed specifications of the routines, and examples of their use.

Chapters 3 and 6 also contain a general introduction to graphics and screen windows, respectively. Here (Chapter 3) we also include a description of how the routines deal with errors.

⟩ 3.2   How errors are dealt with

This section is about errors and how the Toolkit software handles them. If this is the first time that you are reading through the manual and the following information seems on the heavy side, we suggest that you skip this and the next section and continue with the section *Introduction to Level 2 graphics routines*. The Toolkit's examples will show you how to trap errors in situations where that is advisable, and you can refer back to these sections for more detail when you want.

By errors, we mean situations where the routines cannot proceed normally for some reason. Without being exhaustive, typical causes might be:

- the user has passed an unacceptable value or data to a routine;
- an arithmetical error such as division by zero is about to occur;
- the workspace needed by the routine is larger than the free memory remaining (heap full);
- some action by the user (pressing $\boxed{\texttt{Break}}$, for example) forces the routine to stop.

Routines which encounter an error situation use three global variables to pass information back to the calling routine. These variables and their types are:

- **Errorflag**   : Boolean;
- **Errorcode**   : Integer;
- **Errorstring** : Linestring;

These variables are part of the **Cit_core** unit. They are initialised to **false**, 0, and the null string, respectively.

If **Errorflag** is true when a Level 2 Toolkit routine is called then the routine will exit immediately and the values of all three global error variables will remain unchanged. The only exceptions to this rule are the routines in units **Cit_ctrl** and **Cit_menu**, which will continue to work normally even if **Errorflag** is **True**. The reason for these exceptions is that these two units are used to help you deal with errors and must therefore continue to operate after an error is detected. For lower-level routines, the effect of the status of **Errorflag** follows no fixed rules.

If **Errorflag** is false when a Level 2 Toolkit routine is called, then the routine runs normally and the values of all three global error variables will remain unchanged unless an error condition is encountered. If an error is encountered, then **Errorflag** is set **true**, **Errorcode** is set to an integer error code, and **Errorstring** is set to the corresponding error message. A full list of all possible Toolkit error codes and their meanings is given in Appendix 3.

These conventions ensure that once an unexpected situation has arisen, the Level 2 Toolkit routines will have no effect until the user has reset **Errorflag** to **false**, and presumably taken some corrective action as well. It is clearly advisable to test **Errorflag** after any call to a Toolkit routine in which an error can arise, particularly if you want to construct a robust user-interface.

It must be stressed that after detecting an error, **your program**

must reset Errorflag:=false before attempting to call any further Toolkit routines (apart from any in `Cit_ctrl` or `Cit_menu`). If your program includes any kind of menu, then you will probably be using a **REPEAT ... UNTIL** structure in which the call to your menu is embedded, and a good place to reset **Errorflag** is often just after the **REPEAT**.

Another point to note is that whenever you call a routine that looks at the keyboard, a check is made to see if the ⎣Esc⎦ key has been pressed, and if it has, then the global variable **Escape** is set **True**. This is *not* treated as an error condition, so if you want a control loop to exit after escape is pressed, you should test for **Escape** in the closing **UNTIL** statement. There are plenty of instances of this in the examples.

Appendix 3 documents all error codes. Many of these are general and can be generated by several routines. Error states specific to a routine are also described in the documentation for that routine. You will find that some routines can never fail, and that others can only fail if incorrect parameters are passed. If your program guarantees correct parameters, then you can dispense with testing **Errorflag** after the Toolkit routine call.

Examples of suitable code to deal with errors are given in conjunction with examples of how to use the graphics routines.

Errors for Level 1 routines are dealt with in exactly the same way as for Level 2, except for the action on **Errorflag** being **True** as discussed above.

) 3.3   Graphics demonstrations

There are three demonstration programs for the `Cit_draw` unit and they are intended to indicate the range of effects available, but not to cover every feature. Listings for demonstration programs are not printed in the text, but the source code is included with the other examples. At this stage we suggest that you do *not* study the source code of demonstration programs. Later in the chapter there are printed examples (also on disc for ease of running) which show how the routines can be used, and after you have studied those would be the time to look at the demonstration source code as examples of how to combine the graphics routines with the facilities of other units.

⟩ *3.3.1*   *Two-dimensional plotting of $y = f(x)$*

Routines: **Draw_curve1, Sample_function1**
           Draw_data1, Draw_revolution

Graphs with two axes (which we will call x, y plots) are the commonest and most familiar form of graph. In order to plot such a graph you need to specify a function, for example $f(x) = xe^{-x}$, and a range of x over which it should be plotted. This is sufficient information for the Toolkit routines to calculate a range of y-values and so set up appropriate axes. Run program **CHAP3\D-1PLOT2** and select the option **draw** from the bar menu to see this function plotted over the range 0 to 4. (See *Guide to running the software* for detailed instructions on running example programs.) There are other options which allow you to change or edit $f(x)$, alter the x-range, clear the screen, sample the function, and draw a surface of rotation generated from it. The **sample** option demonstrates routines **Sample_function1** and **Draw_data1** working together: the function $f(x)$ is first sampled to create an array of data, which is then passed to **Draw_data1** for plotting. Likewise, **Draw_revolution** does not itself evaluate $f(x)$ but uses the array of sampled data to build a surface of revolution which is then viewed in three dimensions.

⟩ *3.3.2*   *Parametric curves in two or three dimensions*

Routines: **Draw_curve2, Draw_curve3**

In two dimensions the most general form of curve is the parametric form $x = x(s)$, $y = y(s)$ where $x(s)$ and $y(s)$ are functions of s which is called a *parameter*. The curve in the previous subsection can be represented in this way, for example:

$$x = s, \; y = se^{-s}$$

where s will run over the same range as x. As an example of a more complicated curve which cannot be represented in the form $y = f(x)$ consider

$$x = (1 + \cos(3s))\cos(s), \; y = (1 + \cos(3s))\sin(s)$$

You can see what this curve looks like by running **CHAP3\D-2XYZ** for which this is the default curve: simply select the **draw** option.

The parametric form can be extended to three dimensions using $x = x(s)$, $y = y(s)$ and $z = z(s)$, but of course it is not possible to make a true plot of a three-dimensional function on a flat screen and instead one must use representations such as a perspective view. Run **D-2XYZ** again and this time select the **3-D** option, then clear the screen and draw again. You should then see the curve

$$x = (1 + \cos(3s)) \cos(s), \; y = (1 + \cos(3s)) \sin(s), \; z = 4s(1 - s)$$

You can inspect this function from different views using the **view** option.

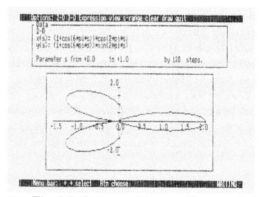

Figure 3.1 The default parametric curve for example **D-2XYZ**

) 3.3.3 *Plotting functions of two variables*

Routines: **Sample_function2, Draw_contours, Draw_surface**

The previous two demonstrations have been concerned with curve plotting, and curves are essentially one-dimensional structures since only one parameter (e.g. x or s) varies over the curve. Here we demonstrate routines for plotting functions of two variables, $f(x, y)$. The two main techniques are to draw perspective views of such functions as surfaces, or to draw contours either flat (as on a map) or in relief.

The routine **Draw_surface** has options to draw contours, surfaces with hidden lines, or wireframes (surfaces without hidden lines). The demonstration program **CHAP3\D-3SURF** shows these techniques in action. When you choose the **contours** option, you specify the number of contours to be plotted. The routines work out the range of heights and hence can calculate a suitable interval between the 'height' associated with each contour.

As was the case with the surface-drawing features in **D-1PLOT2**, the function $f(x, y)$ is first sampled to create an array of data, which is then passed to **Draw_surface** for plotting.

) 3.4   Introduction to Level 2 graphics routines

Level 2 routines require the minimum of information or data to do their work. For example, to plot crosses at a number of points, you need an array with the coordinates of the points, some information about the way that the data are stored in the array, and the colour to be used for the crosses; there is a Level 2 routine to choose and draw suitable axes and plot the crosses.

The *Scientific Programmer's Toolkit* graphics routines support either two-dimensional or three-dimensional graphs. You can plot points or join up a series of points to form a curve, draw surfaces, wireframe diagrams, contours, or bar charts.

) 3.4.1   *Specifying data*

There are two ways of specifying the data to be plotted: by *function* or by *data array*.

- By function specification we mean that there is a Turbo Pascal function defined in the user's program (**f(x)** for example) which can be called to determine the value to be plotted: when you call the plotting routine, you pass a *pointer* to this function so that the plotting routine can call **f(x)** as it sees fit. (Do not worry if you are not familiar with pointers: it is very easy indeed to pass a pointer to a routine and the way that you do this will always be shown in the section on how to call a routine.)
- By data array specification we mean that the user calculates the points to be plotted and stores the values in arrays before calling the plotting routine. When you call the routine, you pass pointers to the data arrays. (Again, passing such pointers is a very easy matter.)

If you have written a function then you could of course evaluate it and store the data in an array, so you may be wondering why we bother to provide a routine to plot functions. There are two reasons: one is that function plotting is very common, so it is convenient to have such routines, and the other is that in order to draw satisfyingly smooth curves it is sometimes necessary to evaluate the function at variable x-spacing which you will not know in advance. Alternatively you may be wondering why we bother to provide routines to plot data arrays, and again there are two reasons: the first is that you will sometimes have data which have not been generated by a function (experimental data, for example), and the second is that you may sometimes wish to generate several columns of results and subsequently wish to plot columns against each other in a variety of combinations, without having to recalculate the data each time. By providing both types of routine we give you more choice. We also provide you with sampling routines that enable you to create data arrays using a function which you have defined.

Some types of plot, namely contouring and surface-drawing, are only provided in a data array specified form. This is for reasons of computational efficiency.

) 3.4.2   *Control of axes and scaling*

All Level 2 routines which carry out plotting use a common convention to determine the way that data should be scaled to fit the current window, and whether or not axes are to be drawn. Routines where this control is provided have an argument **axismode** which is of type **Axismodeoption**, defined in the unit **CIT_DRAW** as:

```
TYPE Axismodeoption = (PRESET, RESCALE,
                       DRAWAXES, AUTOAXES);
```

The effects of each of these settings are as follows:
- **AUTOAXES**: ranges and scales are computed automatically, and appropriate axes are drawn.
- **DRAWAXES**: previously set ranges and scales are left unchanged, but the axes are redrawn as *per* these settings.
- **RESCALE**: ranges and scales are computed automatically, but the axes are not redrawn.
- **PRESET**: previously set ranges, scales and axes are left as they are.

The **DRAWAXES** and **PRESET** options assume ranges and scales have been set previously. This could have been done as described in Chapter 7 (Level 1 graphics), or by a previous call to a Level 2 routine that used the options **AUTOAXES** or **RESCALE**. The **PRESET** option is very useful for superimposing plots of successive functions or data, as long as the first function or set of data sets up ranges into which the subsequent data will fit.

In the rest of this chapter, we will document the two- and three-dimensional function plotting routines first, together with examples, since they are self-contained; then the sampling routines will be documented without examples at that stage, and finally come the routines that plot from data arrays with examples that will also make use of the sampling routines.

⟩ 3.5   The Draw_curve family

These three routines draw paths in two or three dimensions (2-D or 3-D) from a function or functions defined by the user.

Signature:

```
PROCEDURE Draw_curve1 (f        : Funcptr;
                       xl, xh   : Citreal;
                       n        : integer;
                       clr      : Citcolor;
                       axismode : Axismodeoption);
```

This routine draws the curve $y = f(x)$, where the function f is provided by the user. For example:

```
{$F+}
function f(x: Citreal): Citreal;
begin
  f := x*x-1
end;
{$F-}
```

Note that far-calls must be switched on (i.e. **{$F+}**) for the function definition.

The curve drawn is based on $(n - 1)$ line segments over the range **xl, xh**, with step control and breaks at discontinuities. The colour of the plot is set by **clr** and **axismode** acts as described in Section 3.4.2 above.

Signatures:

```
PROCEDURE Draw_curve2 (f          : Funcptr;
                       VAR x, y   : Citreal;
                       sl, sh     : Citreal;
                       n          : integer;
                       clr        : Citcolor;
                       axismode   : Axismodeoption);

PROCEDURE Draw_curve3 (f           : Funcptr;
                       VAR x, y, z : Citreal;
                       sl, sh      : Citreal;
                       n           : integer;
                       clr         : Citcolor;
                       axismode    : Axismodeoption);
```

These routines draw parameterised curves:

$x = x(s)$, $y = y(s)$ in 2-D

$x = x(s)$, $y = y(s)$, $z = z(s)$ in 3-D

The user must write a function **f(s)** which sets the variables **x**, **y**, and (in the 3-D case only) **z**. **x**, **y** *etc.* are variables of type **Citreal** to be declared by the user, which must be global to the user program. They are passed to the draw routine as **VAR** parameters. For example:

```
{$F+}
function f(s: Citreal): Citreal;
begin
    x := (1 + cos(3*s))*cos(s);
    y := (1 + cos(3*s))*sin(s);
    f := 1
end;
{$F-}
```

Note that far-calls must be switched on (i.e. **{$F+}**) for the function definition.

The curve will be composed of $(n - 1)$ line segments over the range (**sl, sh**) although some may not be drawn (subject to the value of **f** as described next). The arguments **clr** and **axismode** act as for **Draw_curve1**.

The value returned by the function **f** has the following effect. When the Toolkit routine calls **f** with the first value of **s**, which will be **sl**, the value of **f** returned is ignored and the routine acts as if **f=0**. The action is that no line segment is drawn but **x,y,**... *etc.* become the current point. If **f** is non-zero then a line segment is drawn joining **x,y,**... to the previous point. This provides a mechanism for drawing a curve with several branches in one call. Another use is to establish axes without drawing, which can be done by a call with **axismode =** **AUTOAXES** and an **f(s)** which always returns zero but visits just two points $(xmin, ymin, \ldots)$ and $(xmax, ymax, \ldots)$.

Error action:

There are two general situations under which these routines will take 'Error action' (see *How errors are dealt with* earlier in the chapter).

First, you may have passed a 'bad value' for an argument: for example, **n** less than 2, or **xh** < **xl**. If your function **f** produces constant values for **x**, **y** or **z** this will similarly cause a **bad range** error if you are asking the routines to work out ranges automatically (**axismode** = **RESCALE** or **AUTOAXES**).

The second error situation is that your function **f(x)** or **f(s)** detects an error, for example an attempt to evaluate $\log(x)$ at $x = 0$. Then it is up to your part of the program to decide what to do, but we recommend that you set **Errorflag:=true** and exit. This will lead to an immediate exit from the Toolkit routine, which will not attempt to plot the associated point on the curve.

Example 1

This shows the simplest use of **Draw_curve1** to plot $y = f(x)$, ignoring any possible errors.

```
{ E-1curv1.pas                              }
{ - plot function y=f(x) - MHB/RDH 14/9/89 }

{ call Toolkit units from Levels 1 & 2 }
USES
   Cit_core, Cit_grap, Cit_draw;

{user-defined function}
{$F+}
   FUNCTION f (x : Citreal) : Citreal;
   BEGIN
     f := x * x - 1
   END;
{$F-}
```

```
BEGIN
  { define graphics port }
  Graph_port (5, 5, 75, 20);
  { plot function }
  Draw_curve1 (@f, - 2, 2, 64, CITLIGHTCYAN,
               AUTOAXES)
END.
```

Example 2

This shows the use of **Draw_curve2** to plot a curve defined parametrically, and including an error trap as a safety net.

```
{ E-2curv2.pas                                          }
{ - parametric plot x=x(s), y=y(s) - RDH 14/9/89 }


PROGRAM Curve2;

  { call Toolkit units from Levels 1 & 2 }
  USES
    Cit_core, Cit_grap, Cit_draw, Cit_ctrl;

  VAR
    { program globals for values of x(s), y(s) }
    xv, yv : Citreal;

  {user-defined function}
{$F+}
  FUNCTION f (s : Citreal) : Citreal;
    VAR
      r : Citreal;
    BEGIN
      r := Exp (- s / (2 * Pi));
      xv := r * Cos (s);
      yv := r * Sin (s);
      f := 1.0
    END;
{$F-}
```

```
BEGIN
  { define graphics port }
  Graph_port (5, 5, 75, 20);
  { plot function }
  Draw_curve2 (@f, xv, yv, 0, 8 * Pi, 128,
    CITWHITE, AUTOAXES);
  IF Errorflag THEN
    Report_error (Errorstring)
END.
```

) 3.6   The Sample_function family

These three routines use a function defined by the user to generate arrays
of data. All such arrays must be declared by the user and have elements
of type `Citreal`.

Signatures:

```
PROCEDURE Sample_function1 (f : Funcptr;
                   xl, xh : Citreal;
                     u, a : Arrayptr;
                  n i, n0 : integer);

PROCEDURE Sample_function2 (f : Funcptr;
             xl, xh, yl, yh : Citreal;
                  u, v, a : Arrayptr;
          ni, nj, n0, na0 : integer);

PROCEDURE Sample_function3 (f : Funcptr;
      xl, xh, yl, yh, zl, zh : Citreal;
               u, v, w, a : Arrayptr;
     ni, nj, nk, n0, na1, na0 : integer);
```

The signatures of the function f provided by the user must be, respectively:
1. function f(x   : Citreal): Citreal;
2. function f(x,y : Citreal): Citreal;
3. function f(x,y,z: Citreal): Citreal;
f must be declared with far-calls switched on, i.e. {$F+}.

1. The function is sampled over a mesh size
$$(\text{xh} - \text{xl})/(\text{ni} - 1)$$
The x-values are placed in an array u[i] and the f-values are placed in an array a[i], both arrays of size at least ni × n0.

2. The function is sampled over a mesh size
$$(\text{xh} - \text{xl})/(\text{nj} - 1), (\text{yh} - \text{yl})/(\text{ni} - 1)$$
The x-, y-values are placed in arrays u[j], v[i] of sizes at least nj × n0, ni × n0, respectively. The sampled data are placed in an array a[i,j] of size at least ni × na0. Note that x corresponds to j and y to i.

3. The function is sampled over a mesh size
$$(\text{xh} - \text{xl})/(\text{nk} - 1), (\text{yh} - \text{yl})/(\text{nj} - 1), (\text{zh} - \text{zl})/(\text{ni} - 1)$$
The x-, y-, z-values are placed in arrays u[k], v[j], w[i] of size at least nk × n0, nj × n0, ni × n0, respectively. The sampled data are placed in an array a[i,j,k] of size at least ni × na1 × na0.

Above, n0 is the row dimension of arrays u,v,w thus allowing the sample points to be stored down columns of a 2-D array if required. If u *etc.* are declared as vectors (i.e. 1-D arrays) then take n0=1. If 'nil' is passed as an array reference for an output data item then the corresponding data are suppressed.

Full working examples of these routines are given with the examples for the **Draw_data** family. Here we give some extracts which demonstrate how the array ordering and sizing works.

Example (i):

User declaration—VAR   u,a: array[1..32]   of   Citreal;
   Typical call—Sample_function1(@f,-10,10,@u,@a,21,1);
will store f(-10), f(-9), ..., f(10) in a[1..21] and -10, -9, ..., 10 in u[1..21].

Example (ii):

User declaration—VAR   a: array[1..32,1..2]   of   Citreal;
   Typical call—Sample_function1(@f,-10,10,@a,@a[1,2],21,2);
will store f(-10),f(-9),...,f(10) in a[1..21,2] and -10, -9, ..., 10 in a[1..21,1].

Example (iii):

User declaration— VAR   a: array[1..32,1..64]   of   Citreal;
   Typical call— Sample_function2(
                    @f,-10,10,-5,5,nil,nil,@a,11,21,0,64);
will store f(-10,-5),f(-10,-4),...,f(-10,5) in a[1..11,1] *etc.*
Note that x is constant down columns of a[i,j].

Error action:

Similar remarks apply to those made for the **Draw_curve** family, except that as the **Sample_function** family do not ever set up axes, no error can occur if the user function **f** generates constant values.

If the global variable **Errorflag** is set true by the user function then the associated data point is not stored and the routine exits.

⟩ 3.7   The Draw_data family

These three routines use data stored in arrays to plot points using symbols . + x, or in the case of linear data the points can be joined with line segments. All such arrays must be declared by the user and have elements of type **Citreal**.

Signatures:

```
PROCEDURE Draw_data1 (u, v : Arrayptr;
                      n, n0 : integer;
                        opt : Drawdataoption;
                        clr : Citcolor;
                   axismode : Axismodeoption);

PROCEDURE Draw_data2 (u, v, a : Arrayptr;
                 ni, nj, n0, na0 : integer;
                        opt : Drawdataoption;
                        clr : Citcolor;
                   axismode : Axismodeoption);

PROCEDURE Draw_data3 (u, v, w : Arrayptr;
                      n, n0 : integer;
                        opt : Drawdataoption;
                        clr : Citcolor;
                   axismode : Axismodeoption);
```

Opt is of type **Drawdataoption** = (Blank,Dot,Plus,Xcross,Join)

The data must be stored as described under the **Sample_function** family above. When a vector is specified you may use a 1-D array u[i] (n0 = 1) or a column of a 2-D array such as a[i,j] (j fixed, n0 = row length). (Column j is indicated by passing @a[1,j].) If a nil pointer is passed for any of the arrays **u**, **v** or **w** then the counter **i** is used instead.

1. A 2-D plot is produced. x, y coordinates are contained in two vectors u[i], v[i] of size at least n × n0, i = 1...n.

2. A 3-D plot is produced. x, y coordinates are contained in two vectors u[j], v[i] of size at least n × n0, j = 1...nj, i = 1...ni, and the z coordinate is in a[i,j] which is at least of size ni × na0. The Join option is treated as if it were blank.

3. A 3-D plot is produced. x, y, z coordinates are contained in three vectors u[i], v[i], w[i] of size at least n × n0, i = 1...n.

If axismode is DRAWAXES or AUTOAXES, then axes are drawn in n dimensions where n = 2, 2, 3, respectively.

Error action:

Similar remarks apply to those made for the Draw_curve family, except that as the Draw_data family do not call a user-defined function, the second kind of error situation cannot arise. However you should note that if any of the arrays passed to a Draw_data routine contains constant values, and the axismode argument is set to recalculate ranges, then an error will be generated as the routines will not be able to set up an axis range for that data.

Example

This shows Draw_data1 being used to give a graph very similar to that obtained in example E-1CURV1 above. The data array is generated using Sample_function1, and the data points themselves are plotted.

```
{ E-3draw.pas                                    }
{ - sample and plot y=f(x) - RDH 14/9/89 }

{ call Toolkit units from Levels 1 & 2 }
USES
  Cit_core, Cit_prim, Cit_wind, Cit_grap,
    Cit_draw;

CONST
  RLX = - 2;
  RHX = 2;
  NDATA = 32;
```

```
VAR
  { for storing sampled points }
  data : ARRAY [1..32, 1..2] OF Citreal;
  { graphics window }
  graph_w : Citwindow;

{user-defined function}
{$F+}
  FUNCTION f (x : Citreal) : Citreal;
  BEGIN
    f := x * x - 1
  END;
{$F-}

BEGIN
  Define_window (graph_w, 5, 6, 75, 23,
    { window position }
    CITLIGHTGRAY, CITBLACK,
    { colours of body    }
    CITWHITE, CITBLACK,
    { ....... of margins }
    PIPYELLOW, AXISRED,
    { ....... of axes    }
    MARGINS, False);

  { switch on graphics mode }
  Graphics_mode;
  { open windows }
  Open_window (graph_w, 'Graph of x^2-1');
  { make window into the graphics window }
  Graph_window (graph_w);

  Sample_function1 (@f, RLX, RHX, @data, @data [1,
    2], NDATA, 2);

  { plot function }
  Draw_data1 (@data, @data [1, 2], NDATA, 2, JOIN,
    CITWHITE, AUTOAXES);

END.
```

⟩ 3.8   The Draw_histogram family

These two routines accept data in the same format as **draw_data1** and **draw_data2**, producing 2-D or 3-D bar charts, respectively. The arguments are almost the same as for the corresponding **draw_data** routine, but **Opt** is replaced by data on the size and placement of the bars. Bar widths *etc.* are specified as a decimal of the interval size in the corresponding direction.

Signatures:

```
PROCEDURE Draw_histogram1 (u, v : Arrayptr;
                           n, n0 : integer;
                     width, disp : Citreal;
                             clr : Citcolor;
                        axismode : Axismodeoption);


PROCEDURE Draw_histogram2 (u, v, a : Arrayptr;
                     ni, nj, n0, na0 : integer;
                      width, depth : Citreal;
                             clr : Citcolor;
                        axismode : Axismodeoption);
```

1. Bars are drawn v[i] high, width*(u[i+1]-u[i]) wide, and the left edge of the bar is at u[i]+disp*width*(u[i+1]-u[i]). Note that disp = −0.5 gives a centred bar.
2. Bars are drawn a[i,j] high, width*(u[j+1]-u[j]) wide, and depth*(v[i+1]-v[i]) deep with the left front edge at (u[i],v[i]).

Error action:

Similar remarks apply to those made for the **Draw_curve** family, except that as the **Draw_histogram** family do not call a user-defined function, the second kind of error situation cannot arise. However you should note that if any of the arrays passed to a **Draw_histogram** routine contains constant values, and the **axismode** argument is set to recalculate ranges, then an error will be generated as the routines will not be able to set up an axis range for that data.

Examples

These examples produce 2-D and 3-D bar charts, respectively.

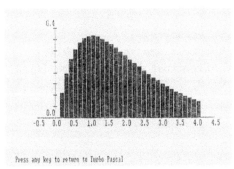

Figure 3.2 The histogram produced by example **E-4HIST1**

```
{ E-4hist1.pas                        }
{ - histogram example -- RDH 11/10/89 }

{ call Toolkit units from Levels 1 & 2 }
USES
   Cit_core, Cit_grap, Cit_draw;

CONST
   RLX = 0;
   RHX = 4;
   NDATA = 32;
VAR
   data : ARRAY [1..32, 1..2] OF Citreal;

{user-defined function}
{$F+}
   FUNCTION f (x : Citreal) : Citreal;
   BEGIN
      f := x * Exp (- x)
   END;
{$F-}
```

```
BEGIN
  { define graphics port }
  Graph_port (5, 5, 75, 20);

  Sample_function1 (@f, RLX, RHX, @data, @data [1,
    2], NDATA, 2);

  { plot histogram }
  Graphparam.cline := CITMAGENTA;
  Draw_histogram1 (@data, @data [1, 2], NDATA, 2,
    1.0, 0, CITWHITE, AUTOAXES);

END.
```

```
  { E-5hist2.pas                              }
  { - 2-D histogram example -- MHB 19/9/89 }

  USES
    Cit_core, Cit_grap, Cit_draw;

  VAR
    a  : ARRAY [1..10, 1..10] OF Citreal;
    u, v : ARRAY [1..10] OF Citreal;
{$F+}
  FUNCTION f (x, y : Citreal) : Citreal;
  BEGIN
    f := Exp (- (x * x + y * y))
  END;
{$F-}

BEGIN

  Sample_function2 (@f, - 2, 2, - 2, 2, @u, @v,
    @a, 10, 10, 1, 10);

  { specify graphics screen }
  Graph_port (1, 1, 80, 25);
  Draw_border;
```

```
WITH Graphparam DO BEGIN
  th := 4 * Pi / 16;
  ph := Pi / 4;
  zflag := True;
END;

Draw_histogram2 (@u, @v, @a, 10, 10, 1, 10, 1,
    1, CITLIGHTMAGENTA, RESCALE)
END.
```

) 3.9   The Draw_surface family

The properties of functions of two variables can be displayed by drawing surfaces based on data points **z** defined on a mesh and stored in 2-D arrays.

Signature:

```
PROCEDURE Draw_contours (u, v, a : Arrayptr;
        n1, nj, n0, na0, ncontours : integer;
                        zl, zh : Citreal;
                           opt : Drawobjectoption;
                           clr : Citcolor;
                       axismode : Axismodeoption);

PROCEDURE Draw_surface (u, v, a : Arrayptr;
               n1, nj, n0, na0 : integer;
                          opt : Drawobjectoption;
                    clr1, clr2 : Citcolor;
                      axismode : Axismodeoption);
```

The **opt** argument of **Draw_contours** may be set to one of CONTOURS, CONTOURS2D and CONTOURS3D; the **opt** argument of **Draw_surface** must be either WIREFRAME or SURFACE.

Contours are usually drawn flat as on a map but they can also be drawn in perspective 3-D by their height. If the CONTOURS option is used, the choice of 2-D or 3-D is made by setting parameters **zflag**, **th** and **ph** in the record **Graphparam** (in Level 1 unit **Cit_grap**). Alternatively, the dimensions used may be forced by using the CONTOURS2D or CONTOURS3D options.

A **Wireframe** is drawn on the mesh points. One can see through the frame. The sides are coloured differently (**clr1** and **clr2**).

When a **Surface** is drawn the hidden lines in the frame are removed. The top-side of the surface is coloured **clr1** and the bottom-side is coloured **clr2**.

The **x**, **y** coordinates of the mesh on which **z** is defined are contained in two vectors **u[j]**, **v[i]** of size at least $n \times n0$, $j = 1 \ldots nj$, $i = 1 \ldots ni$, and the **z** coordinates (heights) are in **a[i,j]** which is at least of size $ni \times na0$. As with the **Draw_data** family **u** and **v** may be columns of a 2-D array, or nil.

When contouring, there are **ncontours** drawn in the range (**zl**, **zh**) at intervals of $(zh - zl)/(ncontours + 1)$. Contours are found by interpolation from the **z**-values and they do not necessarily go through mesh points.

If **zl** = **zh** then the z-range for the contours is computed automatically. If **zl** = **zh** and **ncontours** = 1, then one contour is drawn at height **zh**.

The **axismode** argument has the effect described in Section 3.4.2.

Error action:

Similar remarks apply to those made for the **Draw_curve** family, except that as neither routine calls a user-defined function, the second kind of error situation cannot arise. However you should note that if any of the arrays contain values which are the same at every mesh point, and the **axismode** argument requires ranges to be recalculated, then an error will be generated as the routines will not be able to set up an axis range for those data.

Example

This shows **Draw_surface** being used for 2-D contours, a hidden-line surface, and 3-D contours. The data array is generated using **Sample_function2**.

```
{ E-6surf.pas                                    }
{ 3 ways of plotting f(x,y) - MHB/RDH 14/9/89 }

{ call Toolkit units from Levels 1 & 2 }
USES
   Cit_core, Cit_prim, Cit_grap, Cit_draw;
```

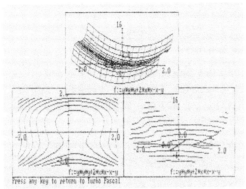

Figure 3.3 The display produced by **E-6SURF**: three ways of representing a function of two variables

```
CONST
  RLX = - 2;
  RHX = 2;
  RLY = - 2;
  RHY = 2;
  NDATA = 16;
  NCONTRS = 12;

{ grids and mesh }
VAR
  data : ARRAY [1..16, 1..16] OF Citreal;
  xmesh, ymesh : ARRAY [1..16] OF Citreal;

{$F+}
  {user-defined function}
  FUNCTION f (x, y : Citreal) : Citreal;
  BEGIN
    f := y * y * y + 2 * x * x - x - y
  END;
{$F-}

BEGIN
  { sample function over grid ... }
  Sample_function2 (@f, RLX, RHX, RLY, RHY,
    @xmesh, @ymesh, @data, NDATA, NDATA, 1,
    NDATA);
```

```
{ specify first graphics screen ... }
Graph_port (1, 12, 40, 24);
Gp_graph_color (CITBROWN);
Draw_border;

{ draw 2d contours ... }
Draw_contours (@xmesh, @ymesh, @data, NDATA,
  NDATA, 1, NDATA, NCONTRS, 0, 0, CONTOURS2D,
  CITCYAN, AUTOAXES);

Graphlabel3 (0, - 2, Graphparam.zlo,
  'f:=y*y*y+2*x*x-x-y');

{ specify second graphics screen ... }
Graph_port (41, 12, 80, 24);
Gp_graph_color (CITBROWN);
Draw_border;

{ define view direction }
WITH Graphparam DO BEGIN
  th := 2 * Pi / 16;
  ph := - 7 * Pi / 16;
  zflag := True;
END;

{ draw 3d contours ... }
Draw_contours (@xmesh, @ymesh, @data, NDATA,
  NDATA, 1, NDATA, NCONTRS, 0, 0, CONTOURS,
  CITCYAN, AUTOAXES);

Graphlabel3 (0, - 2, Graphparam.zlo,
  'f:=y*y*y+2*x*x-x-y');

{ third graphics screen ... }
Graph_port (20, 1, 60, 12);

{ clear area and draw border ... }
Gp_bar (20, 1, 60, 12);
Gp_graph_color (CITBROWN);
Draw_border;

{ draw wireframe ... }
Draw_surface (@xmesh, @ymesh, @data, NDATA,
  NDATA, 1, NDATA, SURFACE,
  CITCYAN, CITBLUE, AUTOAXES);
```

```
Graphlabel3 (0, - 2, Graphparam.zlo,
   'f:=y*y*y+2*x*x-x-y')
```

```
END.
```

) 3.10 Routine Draw_revolution

This routine produces a 3-D view of an object generated by rotating vectors $(r[i], z[i])$ about the z-axis using nslice steps for a full revolution.

Signature:

```
PROCEDURE Draw_revolution (z, r : Arrayptr;
                   n, n0, nslice : integer;
                             opt : Drawobjectoption;
                      clr1, clr2 : Citcolor;
                        axismode : Axismodeoption);
```

The arrays z and r are of size at least $n \times n0$.

The options are CONTOURS, WIREFRAME, SURFACE, SOLID.

- For **Contours** circular contours of z are drawn in colour clr1. clr2 is ignored.
- A **Wireframe** allows all the segments to be seen. The outside is coloured clr1, and the inside is coloured clr2.
- A **Surface** has hidden lines removed from the frame.
- A **Solid** only shows the outside. This is coloured clr1. The parameter clr2 is ignored.

Error action:

Error action is the same as for the Draw_surface family.

Example

This shows a surface which is then rotated.

```
{ E-7revo.pas                              }
{ - Draw surface of revolution - MHB 19/9/89 }
```

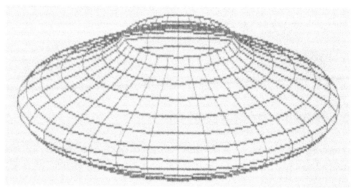

Figure **3.4** One view of the surface produced by **E-7REVO**

```
USES
  Cit_core, Cit_prim, Cit_grap, Cit_draw;

CONST
  NI = 16;

VAR
  a, u : ARRAY [1..NI] OF Citreal;
  k : integer;

{$F+}
  FUNCTION f (s : Citreal) : Citreal;
  BEGIN
    f := 2 + Cos (s)
  END;
{$F-}

  BEGIN
  Sample_function1 (@f, - 3, 3, @u, @a, NI, 1);

  { specify graphics screen }
  Graph_port (1, 1, 80, 25);
  WITH Graphparam DO BEGIN
    th := 5 * Pi / 16;
    ph := 3 * Pi / 16;
    zflag := True;
  END;
```

```
REPEAT
  FOR k := 0 TO 1 DO BEGIN
    Draw_revolution (@u, @a, NI, 1, 32, SURFACE,
      CITCYAN, CITBLUE, RESCALE);
    IF Escape OR Break THEN Exit;
    Gp_active_page (1 - k);
    Gp_visual_page (k);
    Gp_clrscr;
    Rotate_axes (0.05, - 0.2);
  END;
UNTIL Escape

END.
```

⟩ Chapter 4

⟩ Mathematical Routines I

⟩ 4.1   Solution of equations

⟩ 4.1.1   Introduction

Finding roots is a perennial preoccupation of mathematicians. Suppose we have a function $f(x)$, then the problem is to find the values of x which make $f(x) = 0$. The function $f(x)$ could be given by a formula such as

$$f(x) = x^2 - 1 \tag{4.1}$$

or by means of a section of computer program. Some functions have no roots; for example

$$f(x) = x^2 + 1 \tag{4.2}$$

always has a value of at least 1 and is never zero. For other functions several values of x may give zero; the first example above is zero for $x = 1$ and also for $x = -1$. The illustration in the next section shows a function with even more such values.

Both the above examples can be solved without a computer, but they give an idea of the technique which can be used when numerical solution is necessary. The simplest method relies on the fact that for values of x on either side of the root, the values of $f(x)$ will have opposite signs. This is clearly seen in figure 4.1. As a detailed example consider $f(x)$ as defined by equation (4.1) above. At $x = 0.5$, $f(x) = -0.75$, and at $x = 2$, $f(x) = 3$. So somewhere in between there must be a value of x

that makes $f(x) = 0$. The next step is to choose a value of x between
these two; the mid-point is the most efficient, and here it is $x = 1.25$,
giving $f(x) = 0.5625$. This means that the root lies between $x = 0.5$ and
$x = 1.25$. If the value $f(1.25)$ had turned out to be negative, the root
would have been between 1.25 and 2. The process can now be repeated
for the new interval $x = 0.5$ to $x = 1.25$. You can see that the length of
the interval is halved at each stage, so that as the process is repeated the
interval rapidly shrinks, and in fact it can be made as small as required
up to the limits of machine accuracy.

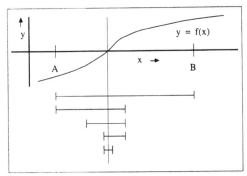

Figure 4.1 Method of bisection: the figure shows a sequence of intervals
homing in on the root of $f(x) = 0$

This technique is known as the *method of bisection*. It is a simple,
reliable method, and very easy to program as shown by the example
below. It is so straightforward that it was felt that a library procedure
would not offer much of an advantage, which is why a bisection routine
has not been included amongst the procedures. The example below is a
program to find the cube root of any number N. First note how this is
turned into a problem of finding zeroes. We want to find a value of x
such that

$$x^3 = N$$

so we take

$$f(x) = x^3 - N$$

This is put in the form of a **Function** statement in the program below:

```
{1-1 bisec.pas                                          }
{ -  simple non-graphic root finder - RDH 16/9/88}
PROGRAM Bisec;

  VAR
    n, a, b, c, fa, fb : real;

  FUNCTION f (x : real) : real;
  BEGIN
    f := x * x * x - n
  END;

BEGIN
  Write ('Find cube root of what number? ');
  Read (n);

  { a, b are numbers which bracket the root }
  { give them initial values .. }
  a := 0;
  b := (n / Abs (n)) * (1 + Abs (n));

  { now find f(a) and f(b) and store .. }
  fa := f (a);
  fb := f (b);

  { now carry out the bisection loop }
  REPEAT
    c := (a + b) / 2;
    {the mid-way value}
    IF f (c) * fa >= 0 THEN
      a := c
    ELSE
      b := c;
  UNTIL Abs (a - b) < 1E-8 + Abs (c * 1E-6);

  { print the result }
  Writeln ('Cube root is ', c:12:6)

END.
```

There is a general point to be learnt from this program. At the end
of the **repeat**... **until** loop there is a test to see whether the values of **a**

and **b** are close enough together for the **repeat** loop to be stopped. The value on the right, `1e-8 + Abs(c*1e-6)`, is called a tolerance. It is made up of two parts: the `1e-8` by itself is the *absolute error tolerance*, while `1e-6` is the *relative error tolerance*. Both are needed: if the absolute tolerance were omitted and the value of **N** chosen was zero or a very small number, the loop might never stop. On the other hand, if the relative part were omitted and the **N** chosen was very large, it would be impossible to get the values of **a** and **b** within `1e-8` of each other because the accuracy of the computer is relative to the size of the numbers being used. The value `c*1e-6` gets larger in proportion to the size of the answer and so takes care of this problem.

The values chosen for the tolerances need to be chosen according to circumstances. You will no doubt have an accuracy in mind according to the use to which the results are to be put, but you will also need to take into account that the computer cannot produce results of arbitrarily high accuracy. The limited accuracy of the machine gives rise to the idea of the *machine epsilon*. Denoted by the variable **eps**, it is the smallest number such that

$$(1 + eps) <> 1$$

to the precision of the computer. It gives a lower limit to the relative precision which can be demanded from any calculation on the machine. It is only necessary to find it to within a factor of 2, something which can be done using this program:

```
{ 1-2 epsilon.pas                        }
{  - find machine epsilon - RDH 16/9/88 }
PROGRAM Epsilon;

  VAR
    eps : real;

BEGIN
  eps := 1.0;
  REPEAT
    eps := eps / 2.0;
  UNTIL 1.0 = 1.0 + eps / 2.0;
  Writeln ('eps = ', eps)
END.
```

On a standard DOS machine using Turbo Pascal type **real**, this gives the result

eps = 1.8E-12 (approximately).

Note that **eps** is a Toolkit global variable which is initialised automatically according to the hardware and software options that apply. In this example the program to find cube roots has not aimed for maximum accuracy, and in practice it is not usual to push accuracy to the limit.

For many practical purposes the method of bisection is adequate. But in more complicated problems, where perhaps the function $f(x)$ is itself the result of an elaborate computation, the method can be too slow. The function ZEROIN overcomes this by beginning with the method of bisection and later switching to more powerful methods. This guarantees that ZEROIN performs at least as well as the method of bisection (measured by the number of function evaluations needed to get an answer) but usually it will be much faster.

It must be pointed out that neither the method of bisection nor ZEROIN will help initially in finding the approximate location of the roots. In practice this is seldom a problem as the user generally has a rough idea of where to look. A thorough discussion of numerical methods for root finding is beyond the scope of this book. The section *Further reading* in the *Introduction* suggests suitable references.

) 4.1.2   *Working example of finding roots*

Filename: **1-3ROOTS.PAS**
To run the working example see *Guide to running the software*.

The default for the function whose roots are to be found is $0.5x - 5\sin x$. The user may change or edit this function using the standard facilities provided by the Toolkit.

The program is operated by selecting options from the menu bar. These should be self-explanatory: there are options for setting the function, setting the parameters needed to sketch the function, *etc.* All input data are requested in verify mode (see *Guide to running the software*).

After sketching the function, select the **root** option and set the search interval from a to b where the values a and b are values of x that bracket the required root. You also need to set a tolerance. The program checks that the function values $f(a)$ and $f(b)$ do indeed have opposite sign; otherwise an error message "**f(a),f(b) not of opposite sign**" is displayed. Note that this message may also occur if a and b bracket more than one root, and in neither case is it sensible to continue. The tolerance controls the accuracy to which the root will be found. A positive value should normally be used, but if 0 is entered the working example attempts to find the root to within 4*(machine epsilon).

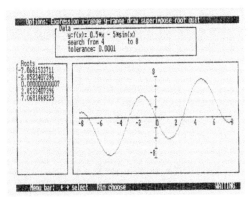

Figure 4.2 A display from program **1-3ROOTS**. All the roots in the range $-8 \leq x \leq 8$ have been located

) *4.1.3   Routine ZEROIN*

Routine name: **ZEROIN**
Kept in unit: **CIT_MATH**
Purpose: Find zero of $f(x)$ in a given range

This routine has been adapted from *Computer Methods for Mathematical Computations* by Forsythe, Malcolm and Moler (Prentice-Hall, 1977).

The unit name **CIT_MATH** must be declared in the **Uses** ... statement at the head of your program.

ZEROIN requires a function to evaluate $f(x)$ which must be declared in the user's program. For example:

```
{$F+}
function f(x : Citreal) : Citreal;
   begin
     f := x*x-5*x+1
   end;
{$F-}
```

The compiler option {$F+} forces 'far-call' compilation which is needed in Turbo Pascal so that ZEROIN can call f(x).

Before calling ZEROIN, the program must decide on two values of x, say a and b, which bracket a root. ZEROIN checks that f(a) and f(b) have opposite sign, and if not it will exit with an error message. The user must also choose a tolerance to control the precision to which the root is found. ZEROIN always converges and is usually much faster than the well-known method of bisection alone.

Signature:

```
function zeroin(a,b,tol: Citreal;
fptr: pointer) : Citreal;
```

Input parameters:

Call:	ZEROIN(a,b,tol,@f)
a,b:	Range of x including a root
tol:	Tolerance for the result (positive or zero)
@f:	Pointer to user's function f(x)

Output parameters:

ZEROIN returns a value of type **Citreal** within **4*eps*x+tol** of the actual root where **eps** is the relative machine precision.

⟩ *4.1.4   Examples of use*

⟩ *Solving a quadratic equation*

Find the roots of the quadratic equation $x^2 - 5x + 1 = 0$.

```
{ 1-4 zer1.pas                      }
{ -  find and print roots - RDH 4/2/89 }
PROGRAM Zeroin1;

  USES
    Cit_core, Cit_math;

  VAR
    f_ptr : Funcptr;
```

```
{$F+}
  {Defines the function whose roots are to be found}
  FUNCTION f (x : Citreal) : Citreal;
  BEGIN
    f := x * x - 5 * x + 1
  END;
{$F-}

BEGIN
  { Set pointer to allow Zeroin to call f(x) }
  f_ptr := @f;

  { We need to know that the
    roots lie between 0,1 and 1,5. }
  Writeln ('First root  = ', Zeroin (0.0, 1.0,
    1E-8, f_ptr):12:8);
  Writeln ('Second root = ', Zeroin (1.0, 5.0,
    1E-8, f_ptr):12:8);
END.
```

Results:

```
First root  = 0.20871215
Second root = 4.79128785
```

) *Suspension bridge height*

The roadway of a suspension bridge is perfectly horizontal, and an engineer is told that if x is the horizontal distance along the roadway, measured from the centre of the span where the cable has its lowest point, then the height of the cable above the roadway is given by

$$y = 5\,cosh(\frac{x}{30})$$

where all distances are in metres and $cosh(x)$ is the hyperbolic cosine function

$$cosh(x) = \frac{1}{2}(e^x + e^{-x})$$

The span is 200 m long, so $-100 \le x \le 100$, and the engineer wants to know to the nearest centimetre the horizontal distance at which the cable height is 35 m, which is about half the maximum height.

```
{ 1-5 zer2.pas                                  }
{ - Suspension bridge height   - RDH 4/2/89 }
PROGRAM Zeroin2;

   USES
     Cit_core, Cit_math;

   VAR
     f_ptr : Funcptr;

   FUNCTION cosh (x : Citreal) : Citreal;
   BEGIN
     cosh := (Exp (x) + Exp (- x)) / 2.0
   END;
{$F+}
   {Defines the function whose roots are to be found}
   FUNCTION f (x : Citreal) : Citreal;
   BEGIN
     f := 5 * cosh (x / 30) - 35.0
   END;
{$F-}

BEGIN

   { Set pointer to allow Zeroin to call f(x) }
   f_ptr := @f;

   Writeln ('Cable height = 35m at x = ',
     Zeroin (10.0, 90.0, 1E-3, f_ptr):9:5);
END.
```

Results:

```
Cable height = 35m at x = 79.01737
```

Note that $x = -79.01...$ is also a solution.

⟩ Further examples

In more complex cases $f(x)$ is found as a result of another calculation. Further examples illustrating this are given in Chapters 4 and 6.

⟩ 4.2   Maxima and minima

⟩ 4.2.1   Introduction

Techniques for finding maxima and minima have many similarities with the problem of finding zeroes of functions, and it may be helpful to refer back to Section 4.1.

The problem is most easily understood graphically. The starting point is the fact that a known function $f(x)$ can be plotted in the form of an x–y graph with $y = f(x)$. This is illustrated in figure 4.4, where you will notice the 'humps' and 'valleys'; these are called *turning points*. The humps are the maxima, the valleys the minima. In this example there is no limit to their number, but only those lying within the ranges of the axes are shown. As it is quite common to have more than one maximum, it would not be correct to define a maximum as the value of x which gives the function its largest value. If however the range of x is restricted to $a < x < b$ where a and b are values on either side of a maximum, then it would normally be true to say that in this range the function takes its largest value at the maximum. However a caution must be given: this is only true if nothing else peculiar happens in the range (although this would be obvious from the graph). Accordingly only *local maxima* are dealt with here, as distinct from the *global maximum* which would be the highest of all the humps. All this of course applies also to the minima.

If there is a simple formula for the function, then the maxima and minima can usually be found by elementary calculus. For example, suppose

$$f(x) = x^2 - x$$

The method is to differentiate the function to get a formula for its slope

$$f'(x) = 2x - 1$$

Maxima or minima will only occur at places where the slope of the graph is zero, so it is now necessary to solve

$$f'(x) = 0$$

In this example there is only one solution, $x=0.5$. With a more complicated function where zeroes cannot be found by algebra, it would be possible to use the root-finding techniques, for example routine `ZEROIN`, given in Section 4.1. Where the formula for $f'(x)$ is available, this is usually a good method.

However, in the most general form of the problem, $f(x)$ may not be known as a formula but only as the result of a computation. A technique is then required for finding local maxima or minima directly from $f(x)$. In fact a technique is only necessary for the minima as the maxima of $f(x)$ occur at precisely the same places as the minima of $-f(x)$.

⟩ 4.2.2   *The golden section search*

The simplest numerical technique for finding minima is known as the *golden section search*. This is similar to the method of bisection described in Section 4.1. The method is illustrated in figure 4.3; two values of x, a and d, are known to bracket a minimum. Two values b and c are then chosen in this range and the function is evaluated to give $f(b)$ and $f(c)$. Then if $f(b) \geq f(c)$, a minimum must lie in the range given by $b \leq x \leq d$; otherwise it lies in the range $a \leq x \leq c$. The process can now be repeated with updated values of a and d. The trick is to choose values for b and c so that in the next stage while one value becomes the end-point of the new range the other value is bound to be at the right place to be one of b and c again. This means that at each stage only one new value of x is needed for a function evaluation. This property can be achieved by choosing the ratio of $(c - a)$ to $(d - a)$ to be the *golden ratio*, giving the method its name. As with the method of bisection, the interval length gets smaller by a constant ratio at each stage, and so can be made as small as required. The golden ratio r satisfies $r = (1 - r)/r$ and will be found to be 0.618034....

The golden section method is simple to program and as with the method of bisection it has not been included in the procedures. The example below may serve as a model; it finds the minimum of $f(x) = x + \frac{1}{x}$.

```
{2-1 gold.pas                                    }
{ - Golden section search   - RDH 7/10/88}
PROGRAM Gold;
```

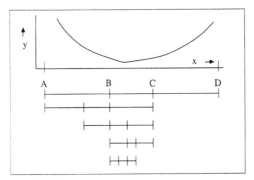

Figure 4.3 Golden section search: the figure shows a sequence of intervals homing in on the minimum of the function $y = f(x)$

```
VAR
  a, b, c, d, fb, fc, gr : real;

FUNCTION f (x : real) : real;
BEGIN
  f := x + 1 / x
END;

PROCEDURE newb;
BEGIN
  d := c;
  c := b;
  fc := fb;
  b := d - gr * (d - a);
  fb := f (b)
END;

PROCEDURE newc;
BEGIN
  a := b;
  b := c;
  fb := fc;
  c := a + gr * (d - a);
  fc := f (c)
END;
```

```
BEGIN
  { set the Golden Ratio }
  gr := (Sqrt (5) - 1.0) / 2.0;

  { assume starting values }
  a := 0.1;
  d := 4.0;

  { calculate starting values for B,C }
  b := d - gr * (d - a);
  c := a + gr * (d - a);
  fb := f (b);
  fc := f (c);

  REPEAT
    IF fb >= fc THEN
        newc
    ELSE
        newb
  UNTIL Abs (b - c) < 1E-8;

  { print the result }
  Writeln ('Minimum found at x = ', b:12:9);
  Writeln ('function value     = ', f (b):12:9)

END.
```

Results:

```
Minimum found at x = 1.00001892
function value     = 2
```

The correct value of x is exactly 1. Many of the remarks on accuracy made in Section 4.1.1 are also applicable here. The limit on the accuracy of the method is determined not by how small the interval can be made, but by how reliably the computer can evaluate the function $f(x)$ at two values close together. It is in the nature of the problem that there is little variation in $f(x)$ near a minimum, as the slope is zero there. Accordingly for the computer there may be a range of x over which $f(x)$ remains constant, or even varies by a very small amount up or down through numerical rounding errors, and then the method may select the wrong search interval. This is why x has been found only to five correct figures, while the function evaluation appears to be precise.

An important point to note in this example is that the function $f(x)$ is singular at $x = 0$, i.e. it is infinite at $x = 0$. It is obviously important that the search interval does not contain any singularities, which is why $A = 0.1$ to start with in the program.

The golden section search may be adequate for many simple cases, but there are more efficient methods, and when the computer takes too long to evaluate the function it will be important to use these. The routine **FMIN** described in Section 4.2.3 begins with the golden section search, and then when it is possible switches to a faster method.

However neither program helps to locate minima in the first place. As with the root-finding problem, this has to be left to the user's part of the program. The section *Further reading* in the *Introduction* suggests suitable references in which a more thorough discussion of techniques for searching for minima or maxima will be found.

⟩ 4.2.3   *Working example of finding minima*

Filename: **2-2MINIM.PAS**
To run the working example see *Guide to running the software.*

The default for the function whose roots are to be found is $0.5x - 5sinx$. The user may change or edit this function using the standard facilities provided by the Toolkit.

The program is operated by selecting options from the menu bar. These should be self-explanatory: there are options for setting the function, setting the parameters needed to sketch the function, *etc.* All input data are requested in verify mode (see *Guide to running the software*).

After sketching the function, select the **minimum** option and set the search interval from a to b where the values a and b are values of x that bracket the required minimum. You also need to set a tolerance. No check is made that a minimum actually lies in the range; if not the program will home in on the end-point with the lesser function value. The error tolerance controls the accuracy of the answer. A positive value should normally be used, but if zero is entered the program attempts to find the minimum to within the least sensible tolerance. This is discussed more fully in the next section.

Note that if you modify $f(x)$ to be the example $f(x)$ used in Section 4.2.2, or to be any other function with a singularity, then take care that your choice of a and b do not bracket the singularity.

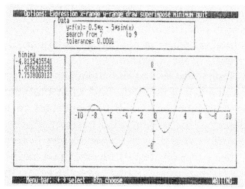

Figure 4.4 A display from program **2-2MINIM**. All the minima in the range $-8 \le x \le 8$ have been located

⟩ *4.2.4   Routine FMIN*

Routine name: **FMIN**
Kept in unit: **CIT_MATH**
Purpose: Find minimum of $f(x)$ in a given range

This routine has been adapted from *Computer Methods for Mathematical Computations* by Forsythe, Malcolm and Moler (Prentice-Hall, 1977).

The unit name **CIT_MATH** must be declared in the **Uses** ... statement at the head of your program.

The method used combines golden section search with parabolic interpolation. It is never much slower than Fibonacci search (which is the theoretical best search technique). If $f(x)$ has a continuous second derivative which is positive at the minimum, the convergence is of order 1.324....

FMIN requires a function to evaluate $f(x)$ which must be declared in the user's program. For example:

```
{$F+}
function f(x : Citreal) : Citreal;
   begin
      f := x + 1/x
   end;
{$F-}
```

The compiler option **{$F+}** forces 'far-call' compilation which is needed in Turbo Pascal so that **FMIN** can call **f(x)**.

Before calling the routine, the program must also decide on two values of x, say a0 and b0, which bracket a minimum. No check that this is true is carried out, and if it is not the routine will home in on the end-point with the lesser function value.

A tolerance (denoted by `tol`) must also be set which helps to control the precision to which the minimum will be found. The function `f(x)` is never evaluated at two points closer than

`eps*ABS(FMIN)+tol/3`

where `eps` = SQRT(m/c *precision*). If the function `f(x)`, as evaluated by the computer, has exactly one minimum when `x` is evaluated at points separated by at least `eps*ABS(x)+tol/3`, FMIN will locate the minimum to within `3*eps*ABS(FMIN)+tol`; otherwise a local minimum is found to the same precision.

Signature:

```
function fmin(a,b,tol: Citreal;
                  fptr: pointer) : Citreal;
```

Input parameters:

Call:	FMIN(a,b,tol,@f)
a,b:	Range of x for the search
tol:	Desired tolerance (greater than zero)
@f:	Pointer to user's function `f(x)`

Output parameters:

FMIN returns a value of type `Citreal`, which is an estimate for the value of x at the minimum, with the precision described above.

⟩ *4.2.5   Examples of use*

⟩ *Minimising a simple function*

Find the miminum of $x + \frac{1}{x}$:

```
{ 2-3 min1.pas                                     }
{ -  find and print a minimum - RDH 5/2/89 }
PROGRAM Fmin1;
```

```
  USES
    Cit_core, Cit_math;
  VAR
    f_ptr : Funcptr;
{$F+}
  {Defines the function whose minima are to be found}
  FUNCTION f (x : Citreal) : Citreal;
  BEGIN
    f := x + 1 / x
  END;
{$F-}

BEGIN

  { Set pointer to allow Zeroin to call f(x) }
  f_ptr := @f;

  { We guess that the minimum
    lies between 0.1 and 10 }
  Writeln ('Minimum is at : ', Fmin (0.1, 10,
    1E-4, f_ptr):12:8);
END.
```

Result:

```
Minimum is at :   0.99999952
```

The correct answer is exactly 1, so the error is under 0.000001, less than the tolerance of 0.0001 which was specified. Note that it is not possible to get a much more accurate answer as the error estimate (see Section 4.2.3) contains a term 3*eps*ABS(FMIN). As eps is approximately SQRT(1E-12) and the function value is 2, this term evaluates to 0.000006 and will dominate the error if the user-set tolerance is reduced much further.

) Spectroscopic data analysis

A chemical analyser produces spectroscopic data in the form of a moving pen trace on paper. The displacement of the pen is proportional to the amount of substance present in the sample. Each substance takes a different time to work through the analyser, and so the time at which a trace reaches its peak helps to identify the substance. However, the response of the analyser is not confined to a single moment as each trace is in the form of a hump which lasts for a certain time. So if for two substances there is only a short interval between the traces, the humps can overlap. In an effort to understand this effect, a chemist wants to find the time at the bottom of the dip between two humps.

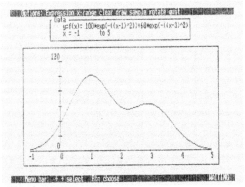

Figure 4.5 The combined effect of two Gaussians, showing the minimum between the two peaks

The humps can be modelled quite well with functions called *Gaussians*. If one substance peaks at time $t=1$ and has strength 100, and another substance peaks at time $t=3$ and has strength 60, the combined response can be given by

$$100\,e^{-(t-1)^2} + 60\,e^{-(t-3)^2}$$

A graph of this function is given in figure 4.5. The following program will find the time of the dip between the peaks:

```
{2-4 min2.pas                                      }
{ - find min between 2 gaussians - RDH 5/2/89}
PROGRAM Fmin2;

   USES
     Cit_core, Cit_math;

   VAR
     f_ptr : Funcptr;
     tmin  : Citreal;
```

```
{$F+}
  {Defines the function whose minima are to be found}
  FUNCTION f (t : Citreal) : Citreal;
  BEGIN
    f := 100 * Exp (- Raise (t - 1,
        2.0)) + 60 * Exp (- Raise (t - 3, 2.0));
  END;
{$F-}

BEGIN
  { Set pointer to allow Zeroin to call f(x) }
  f_ptr := @f;

  { We guess that the minimum
    lies between 1 and 3 }
  tmin := Fmin (1, 3, 1E-4, f_ptr);
  Writeln ('Minimum is at : ', tmin:12:8);
  Writeln ('Function value = ', f (tmin):12:8)
END.
```

Results:

```
      Minimum at t = 2.26166516
      Function value = 55.141603
```

The error tolerance set in this example is certainly too optimistic, and the actual error here will be dominated by the **eps**-dependent term.

⟩ *Further examples*

In more elaborate cases the function is computed as part of the program. Some examples are given in Sections 4.4 and 4.6.

⟩ **4.3   Fitting curves to data**

Experimental data and results from calculations always produce sets of points, but a very common requirement is to 'join up the points' to form a smooth graph. For example, in a simple physics experiment temperature readings might be taken at intervals and then plotted on a graph in order to study the cooling of some object. With x for time and y for temperature the readings are represented by points with coordinates

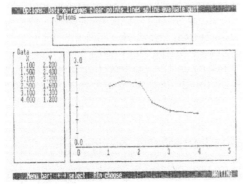

Figure 4.6 Some data points joined with straight lines

(x_1, y_1), (x_2, y_2), *etc.* There is no need for the x-values to be equally
spaced. The simplest way then of joining up the points is to use a
straight line between each pair, as shown in figure 4.6.

If the points are close enough together and the readings accurate,
the straight line method is perfectly satisfactory. It is specially suitable
for plotting functions which can be evaluated quickly and easily, as the
computer can then be programmed to evaluate the function accurately
at points which are close together, and the user can alter the spacing of
the points until a good result is achieved. This is the method used by
the graph-plotting routines in this Toolkit for plotting functions, as in
figures 4.6 and 4.7.

However this technique is often inadequate. If there are a small number of accurate but widely spaced readings, as in the illustration above, a
smooth curve drawn through the points should be more accurate in pre-
dicting intermediate values. The illustration in the next section shows
what can be done. Again if the evaluation by the computer is not fast,
the task could be performed more quickly by obtaining a few widely
spaced results and drawing a smooth curve to find intermediate values.

It is also common, in practice, for many readings not to be accurate.
In this case a smooth curve is needed which may not go through all
or indeed any of the points, but fits the data in an acceptable and
reasonable way.

Methods for fitting curves which go exactly through all the data points
are called *interpolation* methods. One of the most useful of these is
spline interpolation, and the routines in this chapter employ this method.
When the data are known to have errors *best fit* methods are used, and

of these the *least squares* method is the most common. When only a line is required this becomes *least squares regression*. The best fit methods need further mathematical knowledge and are dealt with in Chapter 5, which has examples of fitting data with both lines and curves. The section *Further reading* in the *Introduction* suggests suitable references in which a more thorough discussion of interpolation techniques will be found.

) 4.3.1   *The spline interpolation technique*

The method gets its name from a draughtsman's trick of drawing a smooth curve by flexing a piece of wood or metal, called a spline, so that it runs through all the points. This is fixed with pins and then a pencil is run along its edge to draw the graph. This can be simulated mathematically in several ways. The method described here uses cubic splines, that is each section of the curve is approximated by the formula

$$y(x) = a + bx + cx^2 + dx^3$$

This is a 'cubic polynomial', and the four numbers a, b, c and d are 'coefficients'. They could in fact be chosen to make the cubic go through any four different points, that is, if the points were (x_1, y_1), (x_2, y_2), (x_3, y_3) and (x_4, y_4), it would be possible to choose the coefficients so that

$$y_1 = y(x_1), \ y_2 = y(x_2), \ y_3 = y(x_3), \ y_4 = y(x_4)$$

However if we choose cubics to go through sets of four consecutive points, there is no guarantee that one cubic portion will join up smoothly with the next. The solution is to fit each cubic between two consecutive points only, leaving two free coefficients. This slack, as it were, is used to ensure that the cubics fit smoothly with each other. The joins are called *knots*. This comes from the idea of short flexible overlapping strips tied together with string, each running smoothly into the next, ensuring there are no kinks or corners at the knots. The complete curve with all the cubics determined is called a *cubic spline*.

The procedure **Spline** described in the following sections finds all the coefficients in this way and stores the values in global arrays. There will be four coefficients for each section, and these values are then used by the function **Seval** to evaluate the spline at any required value of x.

The normal technique of evaluation at closely spaced points can then be used to plot the function. Another use for **Seval** might be to locate a crossing point or a maximum.

It is not possible to give a simple estimate of the accuracy of the method as there are an infinite number of smooth curves which can interpolate any given data, and there is no particular reason to suppose that a spline is the 'right' one. The spline itself is found as accurately as computer arithmetic allows, and of course the technique also forces **Seval** to reproduce the data points to machine accuracy.

⟩ *4.3.2   Working example of fitting a spline*

Filename: **3-1SPLINE.PAS**
To run the working example see *Guide to running the software.*

The working example is supplied with arrays dimensioned for up to 20 data points. If more are to be used, the array declarations at the head of the program must be changed, subject to available memory space.

The program is operated by selecting options from the menu bar. These should be self-explanatory: there are options for entering data points, setting the parameters needed to sketch the function *etc.* All input data are requested in verify mode (see *Guide to running the software*).

When the option for data input is selected, you must first set the number of data points, and then the x, y values of each. These are sufficient data to allow the spline coefficients to be computed. The options to display the spline may then be selected. The knots are shown marked with crosses.

At any point $\boxed{\text{Esc}}$ may be used to interrupt the program, returning control to the menu bar.

⟩ *4.3.3   Routines SPLINE and SEVAL*

Routine names:	**Spline, Seval**
Kept in unit:	**CIT_MATH**
Purpose:	**Spline** finds spline coefficients
	Seval evaluates the spline

These routines have been adapted from *Computer Methods for Mathematical Computations* by Forsythe, Malcolm and Moler (Prentice-Hall, 1977).

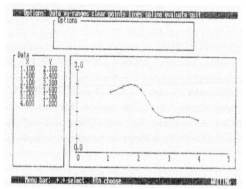

Figure 4.7 The same data points as used for figure 4.6 interpolated with a cubic spline

The unit name CIT_MATH must be declared in the Uses ... statement at the head of your program.

A spline is a sequence of cubic curves, which together make up a smooth curve going through a number of data points (x_i, y_i). Each cubic defines the part of the spline between and including a neighbouring pair of data points, and these cubics are chosen so that they fit smoothly with each other.

Spline fits the spline to the data. Seval evaluates the spline at any point x within the range given by the x-coordinates of the data.

Arrays are needed to store the data and the spline coefficients. These must be declared at the head of the user's program, for example:

```
var
    X, Y, B, C, D : array[1..100] of Citreal;
```

Suppose arrays X[I] and Y[I] are used to hold the data points. The X[I] values need not be evenly spaced, but they must be in ascending order and no two x-values should be equal. The first point is to be stored in X[1], Y[1] and the last in X[N], Y[N]. Because the size of these arrays is not fixed at the time that the Toolkit unit CIT_MATH is compiled, and because of the rules about type compatibility of parameters in procedures, pointers to the arrays are passed as arguments rather than the arrays themselves. This is easily achieved using Pascal's @ operator as shown in the example calls given below.

Signature of Spline:

```
procedure Spline(N : integer;
      xp,yp,bp,cp,dp : Arrayptr)
```

Input parameters for Spline:

Call:	xp:=@X; yp:=@Y; bp:=@B; cp:=@C; dp:=@D;
	Spline(N, xp,yp, bp,cp,dp)
N:	Number of data points
X[1..N]:	x_i-values (in ascending order)
Y[0..N]:	Corresponding y_i-values

Output parameters for Spline:

B[1..N],C[1..N],D[1..N]:   Spline coefficients

The spline is given as follows. Let $z = x - x_i$ and suppose **xx** holds the value of x at which the spline is to be evaluated. Then **Seval(xx)** will return

$$s(x) = y_i + b_i z + c_i z^2 + d_i z^3$$

where i is chosen so that $x_i \le x < x_{i+1}$. At the knots (i.e. the actual data points) the following relations hold:

$$y_i = s(x_i), \; b_i = s'(x_i), \; c_i = \frac{1}{2}s''(x_i), \; d_i = \frac{1}{6}s'''(x_i)$$

Signature of Seval:

```
function Seval(N: integer;
            xx: Citreal;
      xp,yp,bp,cp,dp: Arrayptr) : Citreal;
```

Input parameters for Seval:

Call:	xp:=@X; yp:=@Y; bp:=@B; cp:=@C; dp:=@D;
	Seval(N, xx, xp,yp, bp,cp,dp)
N:	Number of data points
xx:	x-value where spline is to be evaluated
X[1..N],Y[1..N]:	(x_i, y_i)-values as for Spline
B[1..N],C[1..N],	
D[1..N]:	Spline coefficients set by Spline
si:	Starting value for interval search

Output parameters for Seval:

Seval returns the value of the spline at **xx**.

Note that the arrays X[1..N], Y[1..N] which the user's program will have set up before calling **Spline** are not altered by **Spline** and there is no need to reset them before calling **Seval**.

⟩ *4.3.4   Examples of use*

⟩ *Fitting a curve to simple data*

This example produces figure 4.8. It represents the cooling of a mug of hot coffee, a common affliction of computer addicts. Time is plotted in minutes along the x-axis and temperature in centigrade along the y-axis.

```
{ 3-2 spl1.pas                                 }
{ -  simple example using Spline - RDH 14/9/89 }

PROGRAM Spl1;

  USES
    Cit_core, Cit_prim, Cit_text, Cit_wind,
      Cit_grap, Cit_math, Cit_draw, Cit_ctrl,
      Cit_disp;

  { Set up arrays and data }
  CONST
    NDATA = 5;
  TYPE
    data_array = ARRAY [1..NDATA] OF Citreal;
  CONST
    X:data_array = (0.0, 1.0, 3.5, 5.2, 10.0);
    Y:data_array = (90.0, 77.0, 55.0, 45.0, 30.0);

  VAR
    b, c, d : data_array;
    xptr, yptr, bptr, cptr, dptr : Arrayptr;
```

```
  { window variables }
  VAR
    graph_w, results_w : Citwindow;

  VAR
    i : integer;
    xx, result : Citreal;
{$F+}
  FUNCTION s (xx : Citreal) : Citreal;
  BEGIN
    s := Seval (NDATA, xx, xptr, yptr, bptr, cptr,
      dptr)
  END;
{$F-}

BEGIN
  { define working windows }
  Define_window (graph_w, 21, 1, 80, 23, CITWHITE,
    CITBLACK, CITBROWN, CITBLACK, PIPYELLOW,
    AXISRED, MARGINS, False);
  Define_window (results_w, 1, 1, 19, 23,
    CITWHITE, CITBLACK, CITBROWN, CITBLACK,
    CITWHITE, CITRED, MARGINS, False);

  Graphics_mode;

  Open_window (graph_w, '');
  { open windows }
  Open_window (results_w, 'Results');

  Graph_window (graph_w);
  xptr := @X;
  yptr := @Y;
  Draw_data1 (xptr, yptr, NDATA, 1, PLUS,
    CITWHITE, AUTOAXES);

  { Find the spline }
  bptr := @b;
  cptr := @c;
  dptr := @d;
  Spline (NDATA, xptr, yptr, bptr, cptr, dptr);
```

```
{ Evaluate for x from 0 to 10 at 2min intervals }
FOR i := 0 TO NDATA DO BEGIN
  xx := 2 * i;
  Display_real (results_w, 1, i + 1, 5, '', '',
    xx);
  Display_real (results_w, 8, i + 1, 8, '', '',
    s (xx))
END;

{Draw the spline curve s(x) from X[1] to X[Ndata]}
Draw_curve1 (@s, X [1], X [NDATA], 128,
  CITWHITE, PRESET);

Pause;
Quit
END.
```

Results:

0	90
2	66.7326904
4	51.7420487
6	41.338587
8	34.5541118
10	30

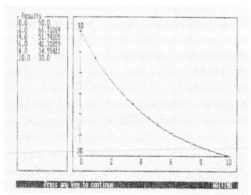

Figure 4.8 The graph resulting from the program **3-2spl1**, showing the cooling of a mug of coffee

) *Estimating peak values from data*

A biological technique for measuring the output of the heart (the rate at which it can pump blood) is to inject a small quantity of dye into a vein, and collect blood samples from an artery. The samples can be analysed for the concentration of the dye, and this can then be plotted as a function of time. It is useful to know the time at which the concentration peaks, though this point may not be explicitly included in the data. One possible technique is to fit a spline, and then use FMIN (see Section 4.2) to locate the peak.

```
{ 3-3 spl2.pas                                }
{ -  Estimate a peak from data - RDH 14/9/89 }

PROGRAM Spl2;

  USES
    Cit_core, Cit_prim, Cit_text, Cit_wind,
      Cit_grap, Cit_math, Cit_draw, Cit_ctrl,
      Cit_disp;

  { Set up arrays and data }
  CONST
    NDATA = 12;
  TYPE
    data_array = ARRAY [1..NDATA] OF Citreal;
  CONST
    X:data_array = (1, 2, 3.0, 4.0, 5, 6, 7, 8, 9,
      10, 11.0, 12);
    Y:data_array = (0, 0, 0.5, 1.1, 12, 18, 22,
      16, 10, 6, 4.5, 3);

  VAR
    b, c, d : data_array;
    xptr, yptr, bptr, cptr, dptr : Arrayptr;

  { window variables }
  VAR
    graph_w, results_w : Citwindow;
```

```
VAR
  i, n : integer;
  tmax : Citreal;
{$F+}
  FUNCTION s (xx : Citreal) : Citreal;
  BEGIN
    s := Seval (NDATA, xx, xptr, yptr, bptr, cptr,
      dptr)
  END;

  FUNCTION minuss (xx : Citreal) : Citreal;
  BEGIN
    minuss := - s (xx)
  END;
{$F-}

BEGIN
  { define working windows }
  Define_window (graph_w, 1, 1, 80, 17, CITWHITE,
    CITBLACK, CITBROWN, CITBLACK, PIPYELLOW,
    AXISRED, MARGINS, False);
  Define_window (results_w, 1, 18, 80, 23,
    CITWHITE, CITBLACK, CITBROWN, CITBLACK,
    CITWHITE, CITBLACK, MARGINS, False);

  Graphics_mode;

  Open_window (graph_w, '');
  { open windows }
  Open_window (results_w, 'Results');

  { Find the spline }
  xptr := @X;
  yptr := @Y;
  bptr := @b;
  cptr := @c;
  dptr := @d;
  Spline (NDATA, xptr, yptr, bptr, cptr, dptr);
  IF Errorflag THEN BEGIN
    Write_string (results_w, 1, 1, Errorstring);
    Exit
  END;
```

```
{ Plot data and spline }
Graph_window (graph_w);
Draw_curve1 (@s, X [1], X [NDATA], 128,
   CITWHITE, AUTOAXES);
Draw_data1 (xptr, yptr, NDATA, 1, PLUS,
   CITWHITE, PRESET);

{ Find the peak }
tmax := Fmin (4, 10, 1E-6, @minuss);
Display_real (results_w, 1, 1, 10,
   'Peak at t = ', '', tmax);
Display_real (results_w, 1, 2, 10, 'Height = ',
   '', s (tmax));
IF Errorflag THEN
   Write_string (results_w, 1, 3, Errorstring);

Pause;
Quit
END.
```

Results:

```
Peak at t = 6.94337391
Height    = 22.0265939
```

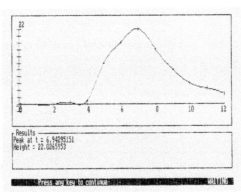

Figure 4.9 The graph resulting from the program **3-3spl2**

This example also demonstrates some of the problems that can occur with data fitting. As the illustration shows, a smooth curve has indeed been fitted, but there is an unexpected dip about $x=3$. This is against intuition, but it simply means that there were not enough data points in the range of x from 4 to 6. Once again the flexible strip gives a good idea of what is happening. If we try to fix such a strip at the crosses, we will find that the amount of bending around $x=4$ will cause the strip to dip. The proper conclusion is that there is not enough data for more than a rough estimate of the peak at $x=7$.

〉 4.4   Integration

〉 4.4.1   Definite and indefinite integrals

With any given function $f(x)$, we can take x–y axes and plot the graph of $y = f(x)$. The area between the curve and the x-axis is the integral of $f(x)$. Where the area is defined between specific values of x, say $x = a$ and $x = b$, it is said to be a *definite integral* on the interval $[a, b]$, with the values a, b as the *limits*. It is usually written in the mathematical form

$$\int_a^b f(x)dx$$

When the limits have not been specified, it is called an *indefinite integral*, written similarly but without the a and b.

If the indefinite integral of $f(x)$ is $F(x)$, then the relation between the definite and indefinite form is

$$\int_a^b f(x)dx = F(b) - F(a)$$

and $f(x)$ is the *derivative* of $F(x)$, written

$$f(x) = \frac{dF(x)}{dx}$$

The notation $F'(x)$ will also be used for the derivative.

With numerical routines, it is not of course possible to obtain the indefinite integral as a formula. The best that can be done is to fix one of the limits, usually the lower limit a, and evaluate the integral for a variety of values of b. This would not be at all efficient with the routines in this section, as the areas overlap and much of the evaluation would be repeated again and again; therefore if the result is required for many values of b, it is better to consider the indefinite integral as the solution of a differential equation, and use the methods given in Section 4.5.

The routines in this section are thus concerned only with definite integrals within limits a and b. These limits must be finite as infinite ranges are not covered. The numerical evaluation of integrals is often called *quadrature*. The section *Further reading* in the *Introduction* suggests suitable references in which a more thorough discussion of numerical integration techniques will be found.

⟩ *Simpson's rule*

A simple method of finding approximate definite integrals is called *Simpson's rule*. It can be applied when the value of the function $f(x)$ is known at every point, or at a and b and at regularly spaced points in between. The interval $[a, b]$ is broken up into so-called panels or sub-intervals. For each panel, let the left side be $x = x1$, the centre $x = x2$, and the right $x = x3$. Then the contribution to the total integral is approximately

$$\frac{1}{3}h(f(x1) + 4f(x2) + f(x3))$$

where the interval size, h, is half the width of the panel. Adding the contributions then gives an estimate for the whole integral. Since each panel shares at least one side with a neighbour, a single evaluation at the shared x-value is sufficient for both panels, as in the following example:

```
{4-1 simps.pas                                          }
{ -  Simpson's Rule to evaluate PI - RDH 23/12/88}
PROGRAM Simpson;

  CONST
    A = 0.0;
    B = 1.0;

  VAR
    h, sum : real;
    npanels, i, weight : integer;
```

```
FUNCTION f (x : real) : real;
BEGIN
  f := 4.0 / (1 + x * x)
END;

BEGIN
  Write ('How many panels? ');
  Read (npanels);

  {Calculate panel half-width}
  h := (B - A) / (2 * npanels);

  {Start with the outer values}
  sum := f (A) + f (B);

  {Loop through all panels}
  {weight alternates 4,2,4,2,...}
  weight := 4;
  FOR i := 1 TO 2 * npanels - 1 DO BEGIN
    sum := sum + weight * f (A + i * h);
    weight := 6 - weight
  END;

  { print the result }
  Writeln ('Result is ', (h * sum / 3.0):15:9);
  Writeln ('whereas PI = ', 4 * ArcTan (1))

END.
How many panels? 2
Result is    3.14156827
whereas PI = 3.1415926536E+00

How many panels? 4
Result is    3.141592502
whereas PI = 3.1415926536E+00
```

This shows that Simpson's rule can give quite accurate results. In fact if you run this program again with eight panels, the result agrees exactly with the computer's value for π, which is as accurate as you can ever expect to get. However, this result is exceptionally good as the integrand $f(x)$ is especially nice and smooth. In practice you do not know the correct answer, and then the usual rule of thumb for judging the accuracy of the result is to run the program a second time with twice as many panels (and the interval size halved) as we did above. We can then compare the results, and usually any figures that agree should be reliable. On this basis, comparing the results above from two and four panels, we could expect 3.1416 to be correct to five figures, while the results from four and eight panels lead to an estimate of 3.141593, correct to seven figures. It would appear that by continuing to halve the interval size the result could be made as accurate as required, but this is not the case as eventually the limited accuracy of the computer's arithmetic becomes important. If the interval size gets too small the error in the results may actually increase through rounding-up errors.

) *Romberg extrapolation*

The property of increasing accuracy with decreasing interval size is exploited by a method which not only compares the accuracy at two stages, but works out the trend and predicts an even better result. Then at the end of the calculation, it gives a result and an error estimate. This is called *Romberg extrapolation*. A routine ROMB is provided in this pack, but as another even more powerful routine is available, the version given here is relatively simple, working to a preset number of interval halvings. For examples, see Section 4.4.4. ROMB is a concise and quick routine, recommended whenever the integrand f(x) appears to be well behaved (but see the note of caution below).

Both Simpson's rule and the Romberg extrapolation use a constant interval size h over the range of integration. This is reasonable if the function is smooth, but sometimes it may vary rapidly in a small range of x, and then slowly and smoothly for the rest of the range. A uniform interval size is then a drawback. The accuracy is determined by the difficult, rapidly varying part, which forces a small interval size and so leads to many unnecessary function evaluations in the smooth part, and perhaps even loss of accuracy. It is possible to divide the integration range into sections with different interval sizes, but it would be more efficient if at each stage the program itself could decide the right interval size to achieve the desired accuracy. These methods are known generally as *adaptive quadrature methods*, and one such called QUANC8 (QUAdrature Newton-Cotes eight panel) is provided. It will also tell you if it cannot achieve the desired accuracy or if there is a value of x at which the integrand has properties which upset the method. The price paid for this level of sophistication is of course an increase in code size and memory space needed.

It is necessary to sound a note of caution here. Even with the most powerful routines, you should not expect to be able to 'plug in' blindly any old integrand and get accurate results. Some problem points will be obvious, for example you cannot expect to be able to integrate $\frac{1}{x}$ from a to b if the range includes $x = 0$. Others will be less obvious, for example $\sqrt{x}$ can be evaluated at $x = 0$ (unlike $\frac{1}{x}$) but if you try to integrate it from $x = 0$ to a positive value, accuracy will be poor as the derivative $\frac{1}{2\sqrt{x}}$ is infinite at $x = 0$. This affects the behaviour of the method although the routine does not use the derivative explicitly. More precise advice on what can go wrong is beyond the scope of this book, and if your results do not appear correct, you should seek expert advice, or consult the text-books. As a general rule, always try to have some idea in advance of the result you expect.

⟩ 4.4.2   *Working examples of integration*

Filename: **4-2QUADR**
To run the working example see *Guide to running the software.*
 The default for the integrand is **exp(-x^2)**. The user may change or edit this expression using the standard facilities provided by the Toolkit. The default range of integration is $[0, 1]$ and these values too can be set by the user. The user may select either or both of the Romberg or **QUANC8** methods to carry out the integration, and the results from both methods are kept on screen so that they may be compared.
 The program is operated by selecting options from the menu bar. These should be self-explanatory: there are options for setting the integrand $f(x)$, limits a and b, parameters for the methods, and the methods themselves. All input data are requested in verify mode (see *Guide to running the software*).
 In order to use the Romberg method the program needs a value for **N**, the number of times to halve the interval (see previous section). The default value of 5 is displayed on the screen and may be changed through one of the options. In order to use the **QUANC8** method, an absolute and a relative tolerance are needed: the default values for each of these are **1E-4** and there is an option to change these. (The concepts of absolute and relative tolerances are explained in Section 4.1.1.)
 When one of the two options **Romberg** or **Quanc8** is selected, the integration is performed and the result displayed, together with an error estimate and the number of function evaluations required. For **Quanc8**,

a quantity called **flag** is also displayed, whose full meaning is described in the next section. If there are no problems, **flag** will be zero.

⟩ 4.4.3   *Routines ROMB and QUANC8*

⟩ *Routine ROMB*

Routine name: **ROMB**
Kept in unit: **CIT_MATH**
Purpose: Numerical integration using
                 fixed step size Romberg method

The unit name **CIT_MATH** must be declared in the **Uses** ... statement at the head of your program.

ROMB requires a function to evaluate $f(x)$ which must be declared in the user's program. For example:

```
{$F+}
function f(x : Citreal) : Citreal;
   begin
     f := Sqrt(x)
   end;
{$F-}
```

The compiler option **{$F+}** forces 'far-call' compilation which is needed in Turbo Pascal so that **ROMB** can call **f(x)**.

The routine integrates $f(x)$ from $x = a$ to $x = b$, the values of a and b being passed as arguments of the function call.

The result is returned as the value of **ROMB**. An error estimate is set in a variable passed as a **VAR** parameter in the function call.

When the routine is invoked, the user supplies a value **n** which is the number of times the interval $[a, b]$ will be halved by the routine; this controls both the accuracy and the number of function evaluations. Usually **n** = 5 will give satisfactory results, but larger values will sometimes be necessary. After a certain point an increase in **n** may result in reduced accuracy through machine rounding errors, but this critical value depends on the problem and must be found by trial and error.

For some problems it may be helpful to divide $[a, b]$ into sub-ranges, as the optimum value for **n** may differ between sub-ranges.

It is assumed that the integrand $f(x)$ and its derivatives up to the nth are finite in $[a, b]$, otherwise accuracy is reduced. For example

$$f(x) = x^{\frac{3}{2}}$$

will give problems if $[a, b]$ includes $x = 0$.

Signature:

```
function romb( a,b: Citreal;
                   n: integer;
                fptr: Funcptr;
          VAR errest: Citreal) : Citreal;
```

Input parameters:

Call:	ROMB(a,b,n,@f,errest)
a, b:	Limits for integration range of x
@f:	Pointer to user's function f(x)

Output parameters:

ROMB returns a value of type **Citreal** which is the routine's estimate of the value of the integral.

errest: Error estimate of the result

) *Routine QUANC8*

Routine name:	QUANC8
Kept in unit:	CIT_MATH
Purpose:	Numerical integration using an adaptive stepsize method

The unit name **CIT_MATH** must be declared in the **Uses** ... statement at the head of your program.

This routine has been adapted from *Computer Methods for Mathematical Computations* by Forsythe, Malcolm and Moler (Prentice-Hall, 1977).

QUANC8 requires a function to evaluate $f(x)$ which must be declared in the user's program. For example:

```
{$F+}
function f(x : Citreal) : Citreal;
   begin
      f := Sqrt(x)
   end;
{$F-}
```

The compiler option {$F+} forces 'far-call' compilation which is needed in Turbo Pascal so that QUANC8 can call f(x).

The routine integrates $f(x)$ from $x = a$ to $x = b$, the values of a and b being passed as arguments of the function call. Two error tolerances aber and reler must also be passed: these determine the accuracy which the routine will try to achieve, and are the same as the tolerances used in the example 4-2QUADR.

The routine is based on the Newton-Cotes eight panel formula, and is self-adapting in that it selects sub-interval sizes to keep errors within acceptable bounds.

The result is returned as the value of QUANC8. This estimate of the integral is designed to satisfy the least stringent error tolerance. Three pieces of information are also returned via VAR parameters in the function call: errest is an estimate of the absolute error; nf is the number of function evaluations used; and flag is an indicator of reliability. flag will be zero if there has been no difficulty; otherwise it will take the form XXX.YYY. The integer part of flag (XXX) is the number of sub-intervals where convergence to the required accuracy did not take place; the fractional part (0.YYY) is the proportion of the range (b-a) left to be integrated when the non-convergence was detected. Even when flag is non-zero, the result can still be meaningful.

As is the case with ROMB and any other general purpose quadrature routine, it is assumed that the integrand $f(x)$ and its derivatives are finite in $[a, b]$, otherwise accuracy is reduced. The variables errest and flag will give warning of any such difficulties.

Signature:

```
function quanc8( a,b,aber,reler: Citreal;
                          fptr: Funcptr;
              VAR errest,flag: Citreal;
                       VAR nf: integer) : Citreal;
```

Input parameters:

Call:	quanc8(a,b,aber,reler,@f,errest,flag,nf)
a, b:	Limits for integration range of x
aber, reler:	Absolute and relative error tolerances required
@f:	Pointer to user's function f(x)

Output parameters:

QUANC8 returns a value of type `Citreal` which is the routine's estimate of the value of the integral designed to satisfy the least stringent error tolerance. It also sets the following VAR parameters:

`errest:`	Error estimate of the result
`flag :`	Reliability indicator (see above—
	`flag` = 0 usually means all is OK)
`nf :`	Number of function evaluations

⟩ 4.4.4   *Examples of use*

⟩ *Evaluating a simple integral using ROMB*

It was remarked earlier that certain integrands can be troublesome; this is an example. The integrand is $\sqrt{x}$, giving the indefinite integral $\frac{2}{3}x^{\frac{3}{2}}$. The range of the integration is from 0 to 1, and so the exact answer should be $\frac{2}{3}$.

```
{ 4-3 rombe.pas                          }
{ -  simple use of ROMBERG - RDH 8/2/89 }
PROGRAM Romberg;

    USES
      Cit_core, Cit_prim, Cit_math;

    CONST
      NHALVE = 5;

    VAR
      result, errest : Citreal;
      fptr : pointer;
      ourerrorstring : Linestring;
```

```
{$F+}
  {Defines the integrand}
  FUNCTION f (x : Citreal) : Citreal;
  BEGIN
    Errorflag := (x < 0.0);
    IF NOT Errorflag THEN
      f := Sqrt (x)
    ELSE
      ourerrorstring := 'Sqrt of negative number'
  END;
{$F-}

BEGIN

  ourerrorstring := ' ';
  fptr := @f;
  {To see error handling, comment out the next line}
  {and un-comment the one after that.               }
  result := Romb (0.0, 1.0, NHALVE, fptr, errest);
  { result:=romb(-1.0, 1.0, Nhalve, fptr, errest); }
  IF NOT Errorflag THEN BEGIN
    Writeln ('Result = ', result:14:10);
    Writeln ('Error estimate = ', errest:5);
    Writeln ('No. of function evaluations = ',
      1 + Raise (2, NHALVE):6:0)
  END
  ELSE BEGIN
    Writeln (Errorstring);
    Writeln (ourerrorstring)
  END;

END.


Result= 0.6662876990
Error estimate= 6.9E-04
```

The error is higher than expected. For comparison with the next example, note that as the interval was halved five times there were 2^5 intervals and $2^5 + 1$ ends, and so **ROMB** needed 33 function evaluations.

⟩ *Evaluating a simple integral using QUANC8*

This is the same integrand, but now we are using **QUANC8** and demanding
high accuracy even though trouble is suspected:

```
{ 4-4 quanc.pas                         }
{ -  simple use of QUANC8 - RDH 8/2/89 }
PROGRAM Quanc;
  USES
    Cit_core, Cit_prim, Cit_math;
  CONST
    RELER = 1E-8;
    ABSER = 0.0;
  VAR
    result, errest, flag : Citreal;
    nf : integer;
    fptr : pointer;
{$F+}
  {Defines the integrand}
  FUNCTION f (x : Citreal) : Citreal;
  BEGIN
    Errorflag := (x < 0.0);
    IF NOT Errorflag THEN
       f := Sqrt (x)
  END;
{$F-}
BEGIN
  fptr := @f;
  result := Quanc8 (0.0, 1.0, ABSER, RELER, fptr,
    errest, flag, nf);
  IF NOT Errorflag THEN BEGIN
    Writeln ('Result = ', result:14:10);
    Writeln ('Error estimate = ', errest:5);
    Writeln ('No. of function evaluations = ',
      nf:6);
    Writeln ('Status flag = ', flag:8:3)
  END
  ELSE
    Writeln (Errorstring)
```

```
END.
```

```
Result =0.666666666
Error estimate: 3.8E-13
No. of evals:    273
Status flag:     0.000
```

QUANC8 was able to deal with this integrand, but only at the expense of a large number of evaluations.

) 4.5   Ordinary differential equations

) 4.5.1   Introduction

Ordinary Differential Equations (ODEs) occur in very many applications of mathematics; in particular they are often used to model processes that change with time. An early version of the theory of ODEs was developed by Sir Isaac Newton: his name for ODEs was *fluxions*. Newton's techniques when applied to his famous Laws of Motion led to spectacular success in modelling the orbits of the planets and other bodies such as comets around the sun.

The solution of a differential equation is always a function of some variable; for example, to predict the position of a comet in space from day to day requires the position to be found as a function of time t. As the comet moves in space it has three coordinates and each will be a function of time, such as $x(t)$, $y(t)$ and $z(t)$. For simplicity, we will start by considering differential equations for a single function $y(t)$, say. The variable t is called the *independent variable*; it will not be time t in all applications but as so many applications do involve time, the variable t will be used for this purpose in this section.

A differential equation is simply a relationship between a function and its derivatives. If that function is $y(t)$, the first derivative will be denoted by either of the expressions

$$y'(t) \quad \text{or} \quad \frac{dy(t)}{dt}$$

and the argument t will sometimes be omitted. Higher derivatives will be denoted y'', y''', etc.

The first derivative y' has a simple geometric interpretation as the slope of the curve $y = y(t)$. If the curve is known, the slope can be calculated. An ODE reverses the problem and asks, if the slope is known, what is the curve? To find a solution, one also needs a starting point, called an *initial condition*.

A simple example of a first-order ODE would be

$$y' = -ky \text{ with initial condition } y = 1 \text{ at } t = 0 \tag{4.3}$$

where k is a constant. This equation expresses for instance the discharge of an electrical capacitor through a resistance. The rate at which charge leaves the capacitor is proportional to the amount in the capacitor. The solution is $y = e^{-kt}$.

A slightly more complicated example expresses the dynamics of simple oscillating systems, as for example a simple pendulum or a weight suspended on the end of a spring. Such systems give rise to the *simple harmonic equation*:

$$y'' + p^2 y = 0 \tag{4.4}$$

where y'' denotes the second derivative of y with respect to t, and p is a constant which depends on the physical properties of the system. This is a second-order ODE as y'' is the highest derivative present. Many other first- and higher-order equations arise in physics, chemistry, engineering, etc.

A higher-order equation can easily be put in the form of a system of first-order equations. For example, in the second-order simple harmonic equation let $z = y'$, and then the equation could be written

$$z' = -p^2 y, \; y' = z \tag{4.5}$$

The important feature is that each of these new equations gives an expression for the first derivative of one of the dependent variables (here they are y and z). This can be generalised for an ODE of any order, leading to the general form:

$$y_i'(t) = f_i(t, y_1, y_2, ..., y_n), i = 1, 2, ..., n \tag{4.6}$$

The dependent variables (the functions to be found) are $y_1, y_2, ..., y_n$. There are n of them, and the *order of the system* is said to be n. The derivatives must be known as functions of t and $y_1, y_2, ..., y_n$. In practice there would also be exactly n initial conditions. For example, with the simple pendulum, $y_1(0)$ might be the initial angle and $y_2(0)$ the initial speed.

Every method for solving an ODE numerically requires a stepsize. We cannot find $y(t)$ for *every* value of the independent variable t; we must be content with results at *discrete values* of t, for example at $t = t_0$, $t_0 + h$, $t_0 + 2h$ etc. The stepsize h need not be constant throughout the calculation.

A simple approach can be developed from the geometrical interpretation. For simplicity, consider a first-order system

$$y' = f(t, y) \tag{4.7}$$

as for example the capacitor problem above, where $f(t, y) = -ky$. Let t0 be the value of t_0 and y0 be the value $y(t_0)$. Then if h is small enough, we can consider the slope to be approximately constant over the step. The value of y at $t = t_0 + h$ could then be taken as

```
y1 = y0 + h*f(t0 , y0).
```

This value can then be used as the starting point for the next step and so on, yielding a series of values of y at time intervals h apart. This can also be done with higher-order systems, and then, for example, the simple harmonic equation gives

```
z1 = z0 - h*(p*p)*y0
y1 = y0 + h*z0.
```

If computers had unlimited speed and precision this method could be made to work with h chosen small enough to give the accuracy required. However as h is made smaller and smaller, the large number of steps required leads to an accumulation of arithmetical errors (also known as *rounding errors*). Further, as each step depends on the previous one and they are all approximate, the approximations are compounded and this can sometimes cause the errors to build up to unacceptable levels, leading to *instability* in the solution. This means that more powerful methods must be found to allow the use of reasonable stepsizes. Indeed the whole art of devising methods for solving ODEs lies in the choice of h to give good accuracy for reasonable stepsizes and stable solutions.

Three routines are provided for solving ODEs. The first, Rk4, is relatively simple, fast and compact, but requires a fixed stepsize and does not of itself give the user any information on the accuracy. It will nonetheless be very useful in demonstrations and for exploring unfamiliar systems of equations. The second routine, **Rkf**, carries out a simple form of step adjustment which keeps the error per step (known as *local error*) within preset limits, whilst the third routine, **Rkf45**, is more powerful still and provides automatic step and error control over a given time interval. These benefits are obtained at the expense of progressively increasing code length and run time.

The first routine, **Rk4**, can be unstable if the stepsize is too large. The instability is a property not only of the stepsize but also of the solution. The method could therefore work quite well up to a certain value of t before becoming unstable. It is thus important to make an informed choice of h. A reliable way to check the behaviour is to start with a large step (say one-fifth of the timescale over which the solution is expected to vary significantly) and solve for an appropriate time, repeat with half the stepsize (and twice as many steps) and then compare the final values of the solution. A similar technique was used to estimate errors in the preceding section of this chapter (Quadrature).

There is an important class of differential equations called *stiff systems*, where two different forms of behaviour come together, one rapidly decaying (transient) and the other long term and varying more slowly. These systems arise, for example, in chemical reaction kinetics. For stability the stepsize must be kept small relative to the timescale, even when the transient solution has decayed. This leads to so many timesteps for the slowly varying solution that it takes too long and arithmetic errors can become significant. **Rkf45** can cope with mildly stiff cases, but in general these routines are inadequate, and specialised methods are needed.

A full discussion of numerical methods for ODEs is beyond the scope of this book: consult *A Simple Introduction to Numerical Analysis* (mentioned in *Further reading* in the *Introduction*).

⟩ 4.5.2   *Working examples of solving ODEs*

Filename: **CHAP4\5-1RK4.PAS**
To run the working example see *Guide to running the software*.

The default ODE is the simple harmonic equation (4.4) with $p = 1$. This order 2 equation is displayed in the form of expressions for y and y' as in equation (4.5) except that y'' is displayed rather than z'. The user may change or edit these expressions as required.

The program is operated by selecting options from the menu bar. This reads

Options:
Order t dt f(t) y(t) y'(t) y' y'' Solve Compare Quit

These are mostly self-explanatory: options Order, t and dt allow you to change the order, initial time and the timestep; y(t) and y'(t) allow you to set initial values; y' and y'' allow the equation to be changed; Solve starts the calculation of the solution; and finally f(t) and Compare allow you to set up any function of t and print its values over the same t-range as the numerical solution. As set up on entry it gives the exact solution $f(t) = sin(t)$ of the default problem so you can see how well or badly the numerical solution has behaved.

The default value of the timestep (0.19635...) may seem a strange choice at first sight, but it is $\frac{\pi}{16}$, and so after 32 timesteps $t = 2\pi$ and the solution should have returned exactly to its initial values.

As the solution is computed, the values are displayed on the screen. When the screen is full, control returns to the Options menu and the values of t, $y(t)$ and $y'(t)$ are updated, so that if Solve or Compare are then selected again, the solution will be continued.

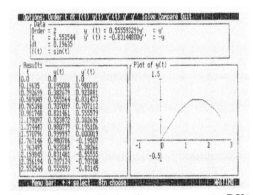

Figure 4.10 A display from program 5-1RK4

) 4.5.3 Routines Rk4, Rkf and Rkf45

) General description

The procedures are designed to solve the ODE system

$$y_i'(t) = f_i(t, y_1, y_2, ..., y_n), i = 1, 2, ..., n \qquad (4.6)$$

The right-hand side of each equation is a function only of the independent variable t and the n dependent variables $y_1, y_2, ..., y_n$, where the initial values are known. Any ODE of order n can be put in this form. To solve the system requires us to find the values of the n y-variables at successive values of t.

The dependent variables are stored in an array. For example,

```
var y: array[1..n] of Citreal;
```

The initial y-values must be stored in this array at the beginning of the calculation. The values will be updated each time a routine is called. The variable used to store t is also passed as an argument to the routines and will also be updated at the end of the step. For **Rk4**, the new value of t is $t+dt$, where dt is the timestep specified when the routine is called. For **Rkf** the new value of t is also $t+dt$ but the value of dt may have been changed by the routine. For **Rkf45**, the new value of t will usually be a specified final value **Tout**, but other actions are possible (see below).

The routines need to evaluate the derivative functions

$$f_i(t, y_1, y_2, ..., y_n)$$

at values of $t, y_1, y_2, ...$ *etc.* which are chosen by the routines themselves; sometimes these values will correspond to the beginning or the end of a step, and sometimes they will be intermediate. In order to allow the routines to do this, the user must write a function to calculate all the f_i, the precise rules for which are about to be given.

These rules may seem complicated but they are quite simple to implement and best understood by examples, of which there are several in the next subsection. (The rules are more complicated than they would be if Turbo Pascal had implemented the ISO Pascal standard which allows a routine to be passed in the argument list to another routine; a full discussion of this problem is given in Appendix 1, under the description of the **Cpas** functions.) The essential point is that the ODE Toolkit routines, **Rk4** for example, need to evaluate all the f_i, as already remarked. The user tells **Rk4** where to find the function that does this by passing a pointer to the function. For example, if the user's function was called **f** then one would pass **@f** in the argument list. Note that **f** has *no arguments*, *even though* **f** *may also need to use the values of* $t, y_1, y_2, ...$ *etc.*

The values of $t, y_1, y_2, \ldots$ will be stored in variables declared by the user: say t and y[j] to be consistent with the declaration above. Since the function f forms part of the user's program, it can have access to these variables. To pass the derivatives back to Rk4, the values calculated must be stored by f in an array yp[j] (say), which must also have been declared by the user. For example, the following lines could form part of a complete program to solve the simple harmonic equation (4.4). In terms of equation (4.5), y[1] is being used to store y, y[2] to store z, yp[1] to store y', and yp[2] to store z'.

```
var
    y,yp: array[1..n] of Citreal;
    t,dt: Citreal:
...
{$F+}
function f: Citreal;
begin
    yp[1] :=  y[2];
    yp[2] := -y[1]
end;
{$F-}
...
{ user sets t,dt, y[1], y[2] prior to call }
Rk4(2,t,dt,@f,@y,@yp)
...
```

The compiler option {$F+} must be switched on for the declaration of f, to force 'far-call' compilation. Note that the value returned as the value of f *is not used*; if you included a statement f:=42 (for example) within the body of f, it would not make any difference.

When f is called (by Rk4 for example), the t-value and y-values at which the derivatives are required will have been stored in the variables t and y[j], j = 0,1, ..., n. f must then evaluate the derivatives using these t- and y-values as arguments for the functions $f_i(t, y_1, y_2, \ldots)$, and store the results in yp[j]. During each timestep, f will be called four or five times in this way, so it is worth giving some thought to writing this code efficiently. Examples are given at the end of the chapter.

The solution values t, y[j] at the end of each timestep are also the initial values for the next step. Therefore once initial values have been set up at the beginning, the routines may be called repeatedly without the user having to alter the values of t or y[j].

The array yp[j] should be treated as workspace: the values in yp[j] are of no particular use to the user.

⟩ *Procedure Rk4*

Routine name:	**Rk4**
Kept in unit:	**Cit_Math**
Purpose:	Take one timestep in the solution of the initial value ODE $y_i' = f_i(t, y_1, y_2, ..., y_n)$ using the fourth-order Runge-Kutta method

The unit name **Cit_Math** must be declared in the **Uses** ... statement at the head of your program.

Rk4 requires a function **f** (say) to evaluate $f_i(t, y_1, y_2, ..., y_n)$ which must be declared in the user's program. See Section 4.5.3 above.

Signature:

```
procedure Rk4( n: integer;
             var t: Citreal;
                 dt: Citreal;
               fPtr: Funcptr;
        yPtr, ypPtr: Arrayptr);
```

Input parameters for Rk4:

Call:	Rk4(n,t,dt,@f,@y,@yp)
n:	Order of the ODE
t:	Value of t at beginning of timestep
dt:	Timestep dt
@f:	Pointer to user's function **f**
@y:	Pointer to array y[1..n] of Citreal
@yp:	Pointer to array yp[1..n] of Citreal

Before the call, t and y[1..n] must be set to the values of t and $y_1, y_2, ..., y_n$ at the start of the timestep. Each time it is called, function **f** must set yp[1..n] to the values of the derivatives, as described in Section 4.5.3.

Output parameters for Rk4:

t and y[1..n] are reset to the values of t and $y_1, y_2, ..., y_n$ at the end of the timestep.

Error indicators for Rk4:

The order **n** must be at least 1. If the user's function **f** sets
Errorflag:=true then the routine aborts.

⟩ *Procedure Rkf*

Routine name: **Rkf**
Kept in unit: **Cit_Math**
Purpose: Take one timestep in the solution of the initial
               value ODE $y'_i = f_i(t, y_1, y_2, ..., y_n)$ using the
               Runge-Kutta-Fehlberg method with a simple step
               adjustment algorithm

The unit name **Cit_Math** must be declared in the **Uses** ... statement
at the head of your program.

Rkf requires a function **f** (say) to evaluate $f_i(t, y_1, y_2, ..., y_n)$ which
must be declared in the user's program. See Section 4.5.3 above.

Signature:

```
Procedure Rkf( n: integer;
       aber,reler: Citreal:
          var t,dt: Citreal;
             dtmin: Citreal;
              fPtr: Funcptr;
      yPtr, ypPtr: Arrayptr;
         var nleft: integer;
        var rkflag: boolean);
```

Input parameters for Rkf:

Call: Rkf(n,aber,reler,t,dt,dtmin,
                             @f,@y,@yp,nleft,rkflag)
n: Order of the ODE
aber: Absolute error tolerance
reler: Relative error tolerance
t: Value of t at beginning of timestep
dt: Current timestep dt
dtmin: Minimum value allowed for timestep **dt**
@f: Pointer to user's function **f**
@y: Pointer to array y[1..n] of **Citreal**

@yp: Pointer to array yp[1..n] of Citreal
nleft: Number of steps to desired staging point of t
                        at current value of dt

Before the call, t and y[1..n] must be set to the values of t and
$y_1, y_2, ..., y_n$ at the start of the timestep. Each time it is called, function
f must set yp[1..n] to the values of the derivatives, as described in
Section 4.5.3.

Output parameters for Rkf:

rkflag: true if step satisfactory, else false
t: Value of t at end of timestep
dt: Current timestep dt
@y: Pointer to updated y[i] values
nleft: Number of steps to desired staging point of t
                        at current value of dt

Error indicators for Rkf:

The order n must be at least 1. If the user's function f sets
Errorflag:=true then the routine aborts.

Before the call, t and y[1..n] must be set to the values of t and
$y_1, y_2, ..., y_n$ at the start of the timestep.

If rkflag is set true, then t and y[1..n] have been reset to the
values of t and $y_1, y_2, ..., y_n$ at the end of the timestep, and dt and
nleft have been adjusted as appropriate.

If rkflag is set false, the error tolerances could not be met without
reducing dt below dtmin.

The algorithm used is:

1. If the error for this step at current dt cannot meet the tolerance
   aber+reler*ABS(y[i]) for all i, halve dt, double nleft and try
   again until the tolerance is met or dt becomes less than dtmin.
2. If dt became too small, do not update the solution. Exit from routine
   returning false.
3. The error tolerance has now been met, so update t and y[i]'s and
   decrement nleft by 1.
4. If estimated error is small enough and nleft is even, double dt and
   halve nleft.
5. Exit from routine returning true.

Note that this algorithm ensures there are always an integral number
of steps left to reach the final value of t. It also ensures that after a
reduction in stepsize an increase cannot be attempted until at least one

more step has been taken; this helps to prevent over-frequent adjustments. Also note that the value of **dt** on exit is not necessarily the value that was used for updating **t**.

) *Procedure Rkf45*

Routine name: **Rkf45**
Kept in unit: **Cit_Math**
Purpose: Advance the solution of the initial value ODE given
               by $y_i' = f_i(t, y_1, y_2, ..., y_n)$ to a given time,
               choosing timesteps to achieve the required accuracy
               and using the Runge-Kutta-Fehlberg method

The unit name **Cit_Math** must be declared in the **Uses** ... statement at the head of your program.

Rkf45 requires a function **f** (say) to evaluate $f_i(t, y_1, y_2, ..., y_n)$ which must be declared in the user's program. See Section 4.5.3 above.

This procedure solves a system of ODEs in non-stiff or mildly stiff cases. The stepsize in the integration is automatically chosen by the routine to meet error criteria set by the user.

The routine has been adapted from FORTRAN subroutine RKF45, written by H A Watts and L F Shampine, Sandia Laboratories, Albuquerque, New Mexico, USA, and published in *Computer Methods for Mathematical Computations* by Forsythe, Malcolm and Moler (Prentice-Hall, 1977).

Signature:

```
procedure Rkf45(    n: integer;
          aber,reler,tout: Citreal:
                   var t: Citreal;
                    fPtr: Funcptr;
            yPtr, ypPtr: Arrayptr
             var rkflag: integer);
```

Input parameters for Rkf45:

Call: Rkf45(n,aber,reler,tout,t,@f,@y,@yp,rkflag)
n: Order of the ODE
aber: Absolute error tolerance
reler: Relative error tolerance

tout:	Required value of **t** at end of step
t:	Starting value of t
@f:	Pointer to user's function **f**
@y:	Pointer to array **y[1..n]** of **Citreal**
@yp:	Pointer to array **yp[1..n]** of **Citreal**
rkflag:	Control code

Before the call, **t** and **y[1..n]** must be set to the values of t and $y_1, y_2, ..., y_n$ at the start of the time interval. Each time it is called, function **f** must set **yp[1..n]** to the values of the derivatives, as described in Section 4.5.3.

Output parameters for Rkf45:

t:	Value of t at end of timestep
@y:	Pointer to updated **y[i]** values
rkflag:	Status code

Before the call, **t** and **y[1..n]** must be set to the values of t and $y_1, y_2, ..., y_n$ at the start of the timestep. **reler** and **aber** must be set to the relative and absolute error tolerances required. **reler** should normally be greater than **100*eps**, but smaller values may be used with care. **aber** can be zero, and the routine will report back if this causes trouble using the status code **rkflag**.

For the first call **rkflag** must be set to 1 or −1. Suppose the step starts with an initial value **t**, and the desired output value is **tout**. If **rkflag** = 1, the routine will normally exit when **t** = **tout**; if **rkflag** = −1, the routine exits after each substep. The status of the calculation can be deduced from the value of **rkflag** on exit (see below). For continuation calls, **rkflag** will normally be left as set by the routine.

The possible values of **rkflag** and their meanings are:

2: successfully reached **tout**.

−2: one substep was successfully made; call again to continue calculation.

3: **reler** was set too small, but has been reset and it is OK to continue.

4: over 500 steps and **tout** not yet reached; it is OK to continue if you wish.

5: solution vanished with **aber** = 0; suggest setting **aber** to non-zero value for one substep (call with **rkflag** = −2).

6: couldn't achieve desired accuracy; must increase **reler** or **aber** before repeat call.

7: Rkf45 seems inefficient here as it would prefer a substep size bigger than **tout** − **t**; the system is stiff.

8: invalid input settings, e.g. $n \leq 0$, or $t \leq$ tout, or rkflag not equal to $+1$ or -1 on input, or reler or aber negative, or rkflag out of range.

Continuation calls of Rkf45:

Normally you would make the first call with rkflag = 1, the routine then returning with rkflag = 2. You can then continue by calling the routine again with the next value of tout. If you used the single-step mode with rkflag = -1, then normally there is no need to reset tout; rkflag = -2 is returned and the routine can be called repeatedly until T = tout. For status codes 3, 4 and 5, after making any changes suggested in the list above, it is possible (if not always sensible) to continue without resetting rkflag. For the remaining codes 6, 7 and 8, after the necessary changes have been made rkflag should be reset to $+2$ (normal mode) or -2 (single substep mode) before continuing.

Error indicators for Rk4:

Most errors and reports for Rkf45 are handled by the status code rkflag: see list above. If the user's function f sets Errorflag:=true then the routine aborts.

) 4.5.4   *Examples of use*

) *The Van der Pol oscillator using Rk4*

This equation

$$y'' + \mu y'(y^2 - 1) + y = 0$$

where μ is a positive constant, is an interesting variation on the simple harmonic oscillator, to which it reverts when $\mu = 0$. The middle term acts to excite oscillations when y is less than 1, and to damp them down when greater than 1. In general there is no closed-form solution of this equation, and it can only be investigated numerically. It is first put into the standard form as shown in the listing, and the solution is then found using Rk4. The constant μ is represented by the program variable mu.

```
{ 5-2vdp1.pas                                    }
{  -  Van der Pol oscillator using Rk4           }
{       (4th order Runge-Kutta) - SMW 2/2/89  }
{                            updated  1/6/89  }
```

```
PROGRAM Vanderpol1;

  USES
    Cit_core, Cit_prim, Cit_math;
  CONST
    ORDER = 2;
    DT    = 0.1;
  VAR
    t    : Citreal;
    y, yp : ARRAY [1..ORDER] OF Citreal;

  { Example function for the second order }
  { Van der Pol non-linear oscillator }
  { Note mu = 0 gives harmonic oscillator }
  { y'' + mu * y' * (y ^ 2 - 1) + y = 0 }
{$F+}
  FUNCTION f : Citreal;
    CONST
      MU    = 0.5;
  BEGIN
    yp [1] := y [2];
    yp [2] :=
      - y [1] - MU * y [2] * (Sqr (y [1]) - 1.0);
    f := 0
  END { f };
{$F-}

  PROCEDURE initialise;
  BEGIN
    y [1] := 0.0;
    y [2] := 1.0;
    t := 0.0
  END { initialise };

  PROCEDURE printheading;
  BEGIN
    Writeln;
    Writeln ('Demo program for Rk4');
    Writeln;
    Writeln ('T', ' ':12, 'Y(0)', ' ':9, 'Y(1)')
  END { printheading };
```

```
    PROCEDURE printresults;

    BEGIN
      Writeln (t:10:8, '    ', y [1]:10:8, '    ',
         y [2]:10:8)
    END { printresults };

  BEGIN
    initialise;
    printheading;
    printresults;
    { loop until T = 1 }
    REPEAT
      Rk4 (ORDER, t, DT, @f, @y, @yp);
      printresults
    UNTIL Abs (t - 1.0) <= 1E-6
  END.
```

T	Y(0)	Y(1)
0	0	1
0.1	0.10236692	1.04592464
0.2	0.208866964	1.08230351
0.3	0.318434186	1.10687978
0.4	0.429770607	1.11729820
0.5	0.541346251	1.11131459
0.6	0.651423267	1.08706321
0.7	0.758108304	1.04334728
0.8	0.859432883	0.97989759
0.9	0.953455308	0.89753466
1.0	1.03837165	0.79818108

⟩ *The Van der Pol oscillator using Rkf*

This is the same problem as solved in the previous section but routine
Rkf is used instead of **Rk4** and the parameter μ is taken as 5 so that
the solution exhibits highly non-linear behaviour. Rather than print out
the entire solution, this example prints results only when the stepsize
changes. There is a region of rapid change between about $t = 1$ and
$t = 3$; notice how the stepsize gets halved as this region is entered and
doubled as it is left.

One property of the Van der Pol oscillator is that after an initial period of non-cyclic behaviour, the solutions settle down to a 'limit cycle'. This can be demonstrated graphically by making minor variations to the following example program; suitable modifications are marked (*...*). With $\mu = 5$ the limit cycle is reached very rapidly. To see an example of slower approach, try $\mu = 1$ with initial values y[1] = 0.1, y[2] = 0.1, and TFINAL = 30.

```
{5-3vdp2.pas                                              }
{- Van der Pol oscillator using Rkf                       }
{    statements in (* ... *) can be added to give         }
{    phase plane plot.                      SMW/RDH 21/9/89}


PROGRAM Vanderpol2;

    USES
      Cit_core, Cit_prim, Cit_math (* , Cit_draw *);

    CONST
      ORDER = 2;
      DTMIN = 0.0001;
      RELER = 1E-5;
      ABER  = 1E-5;

    (* NMAX=500; *)
    VAR
      t, dt, oldstep : Citreal;
      y, yp : ARRAY [1..ORDER] OF Citreal;
      nstepsleft : integer;
      ok : boolean;
    (* results : ARRAY [1..NMAX,1..2] of Citreal; *)
    (* n : integer; *)

    { Example function for the second order }
    { Van der Pol non-linear oscillator }
    { Note mu = 0 gives harmonic oscillator }
    { y'' + mu * y' * (y ^ 2 - 1) + y = 0 }
```

```
{$F+}
  FUNCTION f : Citreal;
    CONST
      MU    = 5;
  BEGIN
    yp [1] := y [2];
    yp [2] :=
      - y [1] - MU * y [2] * (Sqr (y [1]) - 1.0);
    f := 0
  END { f };
{$F-}

  PROCEDURE initialise;

    CONST
      { stopping value for T }
      TFINAL = 5.0 (* 15.0 *);

  BEGIN
    y [1] := - 1.42;
    y [2] := 0.26;
    (* when plotting try also
       y [1] := 0.1;
       y [2] := 0.1;            *)
    t := 0.0;
    dt := 0.2;
    { set number of steps }
    nstepsleft := Round (TFINAL / dt);
    (* n :=0; *)
    { ensure step size divides tfinal }
    dt := TFINAL / nstepsleft
  END { initialise };

  PROCEDURE printheading;

  BEGIN
    Writeln;
    Writeln ('Demo program for Rkf');
    Writeln;
    Writeln ('   T', ' ':11, 'dT', ' ':10, 'y',
       ' ':11, 'y''')
  END { printheading };
```

```
    PROCEDURE printresults;


    BEGIN
      Writeln (t:9:6, '    ', dt:9:6, '    ',
        y [1]:9:6, '    ', y [2]:9:6)
    END { printresults };


  BEGIN
    initialise;
    printheading;
    REPEAT
      oldstep := dt;
      Rkf (ORDER, ABER, RELER, t, dt, DTMIN,
        @f, @y, @yp, nstepsleft, ok);
      (* count and store results *)
      (* n:=n+1;
        results[n,1]:=y[1];
        results[n,2]:=y[2];
        *)
      { print out if stepsize changed }
      IF ok AND (dt <> oldstep) THEN
        printresults
    UNTIL (nstepsleft = 0) OR NOT ok
      (* OR (n=NMAX) *)
      ;
    Writeln;
    Writeln ('Final values:');
    printresults;
  (*
    Writeln;
    Writeln(n,' steps.');
    Writeln('Press RETURN to see phase plane plot');
    Readln;
    graphics_mode;
    draw_data1(@results,@results[1,2],n,2,join,
                                 CITGREEN,AUTOAXES)
    *)
  END.
```

```
Demo program for Rkf
     T            dT           y            y'
  0.100000     0.100000    -1.393474     0.270842
  1.450000     0.050000    -0.691487     1.352541
  1.575000     0.025000    -0.480352     2.116105
  1.862500     0.012500     0.782687     7.442874
  2.131250     0.006250     1.990946     0.706576
  2.237500     0.012500     2.021260     0.034394
  2.425000     0.025000     2.006792    -0.122544
  2.650000     0.050000     1.977399    -0.134547
  2.800000     0.100000     1.957017    -0.137148
  Final values:
  5.000000     0.100000     1.600567    -0.198512
```

It is also possible to plot simultaneous graphs of y against t and y' against y (known as a *phase-plane plot*). This requires the use of Level 1 graphics routines, and an example is given in Chapter 7, program 4-TWIN.

) *Planetary orbit using Rkf45*

In honour of Sir Isaac Newton, example 5-4ISAAC is a simulation of the motion of a planet around its sun using Rkf45. When running, the orbit is shown graphically and simultaneously the position and time are shown numerically. The meaning of Flag is described in the description of Rkf45 given earlier. Unless interrupted the simulation runs for 8 time units which is the analytical result for the planet's 'year' in this case, so that x and y should have returned to their initial values. The power of the method is demonstrated by the fact that this can be seen to be true to nine-figure accuracy.

The example is not intended as a polished orbit solver but as a rather more elaborate programming example than some of the other programs, and because of its length the listing is not printed here.

) *Foxes and rabbits population model*

Cyclic variations in populations of interdependent species can be presented in a mathematical model. Such systems of equations, like the pair in this example, are called *Volterra systems*. The equations are included in the comments in the program listing of **5-6fox2** shown below, but first run **5-5fox1** (listing not given) which solves the equations and draws a graph of the results, which are interpreted as follows. The predator foxes will decrease in numbers unless there are sufficient rabbits available for food. The rabbits in turn can breed quite happily on their own, but when there are too many foxes their numbers decline. Thus the rabbits increase only while the foxes are scarce, and there is a delay before the greater rabbit population leads to a greater number of foxes. The foxes then over-hunt the rabbits, causing a food shortage and again a decline in fox numbers, and thus we have the cyclic change. Fox numbers are plotted horizontally and rabbits vertically; both numbers can be taken in units of 1000.

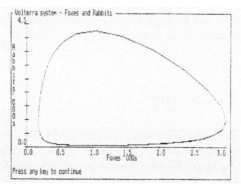

Figure 4.11 The graph resulting from the Foxes and Rabbits example program **5-5fox1**. Fox numbers are plotted along the *x*-axis, rabbit numbers along the *y*-axis

) *Finding the least number of foxes*

Looking at the previous example, it could be of some interest—especially to rabbits—to know when the fox population is at its lowest. This example shows how **Fmin** (see Section 4.2) could be used in combination with **Rk4** (**Rkf** could be used instead with very minor modifications).

The same technique can of course be used to find any of the turning points.

```
{ 5-6fox2.pas                                 }
{  - Foxes and rabbits population modelled    }
{     by Volterra system using Rk4            }
{  - Find least foxes using fmin - SMW 2/2/89 }
{                          updated 1/6/89 }

PROGRAM Fox2;

  USES
    Cit_core, Cit_prim, Cit_math;

  CONST
    ORDER     = 2;
    DT        = 0.1;
    TOLERANCE = 1E-3;

  VAR
    t, told, tmin, oldfox : Citreal;
    y, yp, yold : ARRAY [1..ORDER] OF Citreal;

{$F+}

  { d (foxes) / dt =
    -foxes + foxes * rabbits
    d (rabbits) / dt =
    2 (rabbits - foxes * rabbits)

    y[1] is foxes, y[2] is rabbits }

  FUNCTION f : Citreal;
  BEGIN
    yp [1] := - y [1] + y [1] * y [2];
    yp [2] := 2 * (y [2] - y [1] * y [2]);
    f := 0.0
  END { f };
{$F-}

{$F+}
```

```
{ called by fmin to determine
  number of foxes at time tm }


FUNCTION mf (tm : Citreal) : Citreal;

  VAR
    mft : Citreal;

BEGIN
  mft := told;
  y := yold;
  Rk4 (ORDER, mft, tm - told, @f, @y, @yp);
  mf := y [1]
END { mf };
{$F-}


PROCEDURE initialise;


BEGIN
  t := 0.0;
  y [1] := 3.0;
  y [2] := 1.0
END { initialise };


PROCEDURE printresults;


BEGIN
  Writeln ('Least foxes at time ', tmin:10:8);
  Writeln ('There were ', y [1]:10:8,
    ' x 1000 foxes');
  Writeln ('and ', y [2]:10:8,
    ' x 1000 rabbits')
END { printresults };
```

```
{ main program }
BEGIN
  initialise;
  { repeat until foxes start to increase, }
  { then record values }
  REPEAT
    oldfox := y [1];
    Rk4 (ORDER, t, DT, @f, @y, @yp)
  UNTIL y [1] >= oldfox;
  told := t;
  yold := y;
  { increase started in interval dt before t=told, }
  { so use that as new starting point and use fmin }
  { to locate minimum }
  tmin := Fmin (told - DT, told, TOLERANCE, @mf);
  printresults
END.
```

```
Least foxes at time 3.80047973
There were 0.17869150 x 1000 foxes
and 1.04381783 x 1000 rabbits.
```

Although the computer has printed out the results to nine figures, there is no reason to believe they are all accurate. The error tolerance set for **Fmin** will only be achieved if **Rk4** has got the solution right to a similar accuracy. A check can be made by repeating the calculation as usual with half the stepsize, or by using **Rkf45**.

) *Flight time of a golf ball*

This example shows a combination of **Rk4** and **Zeroin** (see Section 4.1). **Rk4** allows the ball to be followed at regular timesteps **dt**. We can then use **Zeroin** to find more accurately within a timestep the time the ball hits the ground.

```
{ 5-7golf.pas                                    }
{  - Flight time of a golf ball using            }
{    Rk4 (fourth order Runge-Kutta)              }
{  - Find earth contact using zeroin - SMW 2/2/89 }
{                               updated 1/6/89 }
```

```
{
  Ball is launched at velocity VELOCITY at an angle
  THETA. G is acceleration due to gravity, K is the
  drag coefficient, with drag = K * VELOCITY squared.
  Units are metres and seconds.

  y[1] = x, y[2] = y, y[3] = x', y[4] = y'
}


PROGRAM Golf;

  USES
    Cit_core, Cit_prim, Cit_math;

  CONST
    ORDER     = 4;
    DT        = 0.25;
    TOLERANCE = 1E-5;
    G         = 9.81;
    K         = 0.01;
    VELOCITY  = 70.0;
    THETA     = 45.0;

  VAR
    t, t2, tzero : Citreal;
    y, y2, yp : ARRAY [1..ORDER] OF Citreal;

{$F+}

  { Equations are:
    x'' = -k v x'
    y'' = -g - k v y'
    v   = Sqrt(x'^2 + y'^2) }

  FUNCTION f : Citreal;
    VAR
      v : Citreal;
```

```
   BEGIN
     v := Sqrt (Sqr (y [3]) + Sqr (y [4]));
     yp [1] := y [3];
     yp [2] := y [4];
     yp [3] := - K * v * y [3];
     yp [4] := - G - K * v * y [4];
     f := 0.0
   END { f };
{$F-}

{$F+}

{ called by zeroin to determine height at time tz }

   FUNCTION zf (tz : Citreal) : Citreal;

     VAR
       zft : Citreal;

   BEGIN
     zft := t2;
     y := y2;
     Rk4 (ORDER, zft, tz - t2, @f, @y, @yp);
     zf := y [2]
   END { zf };
{$F-}

   FUNCTION radian (degrees : Citreal) : Citreal;

   BEGIN
     radian := Pi / 180.0 * degrees
   END { radian };

   PROCEDURE initialise;

   BEGIN
     t := 0.0;
     y [1] := 0.0;
     y [2] := 0.0;
     y [3] := VELOCITY * Cos (radian (THETA));
     y [4] := VELOCITY * Sin (radian (THETA))
   END { initialise };

   PROCEDURE printresults;
```

```
      BEGIN
        Writeln ('Ball hits ground after ',
          tzero:12:7, ' seconds');
        Writeln ('Distance travelled was ',
          y [1]:12:7, ' metres');
        Writeln ('Horizontal velocity was ',
          y [3]:12:7, ' m/s');
        Writeln ('Vertical velocity was ',
          y [4]:12:7, ' m/s')
      END { printresults };

    { main program }
    BEGIN
      initialise;
      { follow the motion until the height y
        (stored as y[2]) goes negative. Record values. }
      REPEAT
        Rk4 (ORDER, t, DT, @f, @y, @yp)
      UNTIL y [2] <= 0.0;
      t2 := t;
      y2 := y;
      { Ball landed in interval dt before t = t2. So use
        that as new starting point and use zeroin to make
        height zero. }
      tzero := Zeroin (t2 - DT, t2, TOLERANCE, @zf);
      printresults
    END.
```

```
Ball  hits ground after    6.3414030 secs.
Distance  travelled was  127.9410741 metres.
Horizontal velocity was    8.9633988 m/s
Vertical   velocity was  -24.2342346 m/s.
```

As usual, the accuracy can be tested by repeating the calculation with half the timestep.

) 4.6   Fourier transforms

Fourier transforms are a way of analysing signals to determine the relative strength of any frequency. A rainbow demonstrates this, since light is made up of a spectrum of frequencies and the brightness of each band indicates the strength of that particular wavelength in the original beam of light that has been spread out in the formation of the rainbow. Similarly in sound, even a single musical instrument playing a single note does not produce a pure tone as there are always a number of harmonics. By using a microphone, any sound can be converted into a measurable electrical signal which can be recorded. For analysis, the signal is first digitised by measuring the strength at regular intervals. This process is called 'sampling the signal'. The readings may be plotted against time as points, which may be 'joined up' in the manner described in Section 4.3 to produce a graph made up of straight line sections. Provided that the sampling interval is small enough a reasonable representation of the original signal can be achieved. Plotting straight line sections is equivalent to assuming that the signal strength varies uniformly between the sample points. This is the way that modern digital recording techniques work: the recording is a collection of samples which on playback cause the loudspeaker cones to move and reproduce the original air vibrations. The reason for sampling is the familiar one that computers do not have infinite memory, and so a continuous signal must be broken up into a number of discrete steps. Fourier transforms have obvious importance in electronic and sound engineering, but they are also used as a general mathematical technique.

⟩ 4.6.1 Discrete Fourier transforms

Because of the sampling, it is not possible to find the exact Fourier transform; the best that can be done is to find the Discrete Fourier Transform, or DFT for short.

Suppose the signal is given by a function $x(t)$. If the sampling was at N equal intervals of time dt, the sample values would be

$$X_r = x(r\,dt) \quad r = 0, 1, ..., N - 1 \tag{4.8}$$

where the range of t is from 0 to $T = N\,dt$. However it is sometimes required to sample over a range with $t = 0$ in the centre. The first half of the values are then $X_r, r = 0, 1, ..., (\frac{N}{2} - 1)$ as before (N is conveniently taken as even) while the other half are given by

$$X_r = X((r - N)dt), \quad r = \frac{N}{2}, 1 + \frac{N}{2}, ..., N - 1 \tag{4.9}$$

The DFT of these values is defined in terms of the complex exponential function which has the property

$$e^{i\theta} = cos(\theta) + isin(\theta) \tag{4.10}$$

where i denotes the square root of -1. Pure harmonics are described by sines and cosines. The DFT of the sample values is then given by

$$XT_s = \sum_{s=0}^{N-1} X_r \, exp\left(\frac{-2\pi irs}{N}\right) \tag{4.11}$$

where the summation over r is carried out for each of the N values of s from 0 to $N - 1$. Each calculated XT_s will then be a complex number representing the strength and phase of a particular harmonic. The original sample values can be resynthesised by the following formula:

$$XTT_r = \frac{1}{N}\sum_{s=0}^{N-1} XT_s \, exp\left(\frac{+2\pi irs}{N}\right) \tag{4.12}$$

The exponential in (4.12) may be interpreted as follows in terms of harmonics.

The N sample values give N intervals within the total time T, and so $dt = \frac{T}{N}$. If t_r is the time associated with the sample value X_r we then have

$$t_r = r\,dt = r\frac{T}{N} \tag{4.13}$$

The exponential in (4.12) can therefore be written

$$exp(+2\pi it_r f_s) \text{ where } f_s = \frac{s}{T} \tag{4.14}$$

From (4.10), the exponential in (4.14) consists of sines and cosines of $2\pi ft$. Considered as functions of t, these have period $\frac{1}{f}$ and frequency f, and so the frequency associated with XT_s is f_s. However we must carefully consider the range of f for which this interpretation is valid. As the sines and cosines of the complex exponential are cyclic, the XT_s are also cyclic, repeating themselves every N values, that is XT_s is the same as $XT_{(s+N)}$. XTT_r has of course a similar property, and also has only N distinct values.

As the DFT provides N distinct frequency components from N sample values, it should be expected that just as there was a range of T for the sampling, so there should be a range of frequency in the transforms. From (4.13), the interval between each frequency is $df = \frac{1}{T}$, so the range of frequencies is $F = Ndf = \frac{N}{T}$. It turns out that the appropriate range of f is from $-\frac{F}{2}$ to $+\frac{F}{2}$. As XT_s is cyclic, the value at $s = \frac{N}{2}$ equals the value at $s = -\frac{N}{2}$ which corresponds to $f = -\frac{F}{2}$. The values of s from $\frac{N}{2}$ to $N - 1$ should then be interpreted as belonging to the frequencies $-\frac{F}{2}$ to $-df$. Negative frequencies may seem rather strange, but they arise simply because the complex XT_s-values contain both amplitude and phase information.

Notice that the only difference in form between (4.11) and (4.12) is the sign under the exponential and the factor $\frac{1}{N}$ outside the summation in (4.12). The values of XTT_s are of course equal to the original sample values.

The routine FFT can be used to evaluate either (4.11) or (4.12). There is no particular numerical difficulty involved, except that if done in the obvious way each transform would take about N^2 operations, and for large values of N this could be very time consuming. FFT is therefore based on the Fast Fourier Transform algorithm, for which N has to be a power of 2. The number of operations is then proportional to $Nlog(N)$, which is very much faster.

) 4.6.2 Working examples of DFTs

Filename: **6-1DFT**
To run the working example see *Guide to running the software*.

Program **6-1DFT** calculates sample values of a function $x(t)$, allows them to be plotted or printed, takes their DFT and allows that to be plotted or printed too.

The working example is supplied with a default function which may be changed using the standard facilities provided by the Toolkit. The default function is defined as follows:

$$x(t) = \begin{cases} 0 & \text{if } |t| > 1 \\ 0.5 & \text{if } |t| = 1 \\ 1 & \text{if } |t| < 1 \end{cases}$$

In order to give a simple but reasonably general format which allows the definition to be edited with ease, this definition is displayed on the screen in the form:

```
-inf < t < -1        x(t) = 0
-1   < t <  0        x(t) = 1
 0   < t <  1        x(t) = 1
 1   < t < +inf      x(t) = 0
```

In the above definition the t-values -1, 0 and 1 are *junction points* which separate the four intervals of t over which $x(t)$ is defined. The junction points can be changed. The left-most and right-most values (-1 and 1 in this case) are called *cutoffs*, and the middle point is simply called the *junction*. For t-values less than the left cutoff or greater than the right cutoff, $x(t)$ will always be set to zero. (This is a feature of this working example rather than of the DFT method, although it is generally true that it only makes sense to attempt to find the Fourier transform of a function which tends to zero as $t \to \pm\infty$.) In between the cutoffs, the program allows two expressions for $x(t)$ to be set; in the example above they are both 1. As we have described the on-screen definition of $x(t)$ so far, it corresponds to the mathematical definition $x(t) = 0$ if $|t| > 1$, $x(t) = 1$ if $|t| < 1$. This appears to leave out the definition of $x(t)$ at $t = \pm 1$, but in fact in this program the values at the junction points are calculated automatically as the average of the $x(t)$-values marginally on either side of the junction. In our present example, this produces $x(t) = 0.5$ at the cutoffs and $x(t) = 1$ at the junction $t = 0$.

To recap, the program allows editing of (a) the values of the cutoffs and the junction, and (b) the expressions for $x(t)$ for the t-intervals on either side of the junction. This is adequate for many of the simple examples mentioned in text-books on Fourier Analysis, whilst leaving the code of the example program relatively uncomplicated.

Incidentally the method of calculating values at junction points illustrates the correct way of dealing with functions that have discontinuities (jumps) in their values: the value at the jump should be set to the average of the value on either side.

The program is driven with the usual menu bar, which shows:
Options: Expressions M,T function transform quit

The **Expressions** option produces the function definition display discussed above, together with a drop-down menu which allows an item to be selected for editing. The **M,T** option allows the sample size N (recall $N = 2^M$) and the range of t to be changed: the function $x(t)$ is sampled at N equally spaced points throughout the interval $-\frac{T}{2} < t < \frac{T}{2}$. The **function** option produces a drop-down menu which allows the axes ranges for the graph of the function to be changed or the actual sample values to be displayed. The **transform** option gives similar facilities for the DFT of $x(t)$, which is automatically recalculated if the definition of $x(t)$ or the values **M,T** have been changed.

Note that the sample values are stored as complex numbers, and for the usual case of a real function the sample values give the real parts while the imaginary parts are zero. On the graphs, real parts are plotted with solid lines, imaginary parts with dotted lines.

Although **6-1DFT** is intended chiefly to demonstrate the routine **FFT** in action, it is a rather more elaborate program than the working examples of previous sections, and may be useful as an example program, demonstrating how to combine text, graphics, and various sorts of menu.

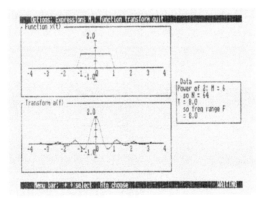

Figure 4.12 Fourier transforms: display resulting from program **6-1DFT**

⟩ *4.6.3 Procedure FFT*

Routine name:	**FFT**
Kept in unit:	**CIT_MATH**
Purpose:	To compute a Discrete Fourier Transform using the Fast Fourier Transform algorithm

The unit name CIT_MATH must be declared in the **Uses** ... statement at the head of your program.

The routine evaluates equation (4.11) from Section 4.6.1:

$$XT_s = \sum_{s=0}^{N-1} X_r \, exp\left(\frac{-2\pi i r s}{N}\right) \qquad (4.11)$$

for $s = 0, 1, ..., N - 1$. N is the number of sample points; this must be an integer power of 2. The values X_r are the sample values, and XT_s is the DFT. Both sets of values are treated as complex and stored in an array F[i,j] where j = 0 for real parts, j = 1 for imaginary parts; i runs from 0 to N − 1, as do subscripts r and s in (4.11) above. A suitable array declaration for N up to 256 would be:

VAR F : array[0..255,0..1] of Citreal;

The routine operates directly on the array elements; before calling the routine the user must store the sample values in F[i,j], and on return F[i,j] holds the result, so the original values are lost. To allow inverse transforms, FFT will also evaluate (4.11) but with a + sign under the exponential.

Signature:

```
procedure FFT( Fptr : Arrayptr;
                  M : integer;
               sign : Citreal)
```

Input parameters:

Call:	FFT(@F,M,sign)
@F:	Pointer to user's array F[i,j] which must contain the sample values on entry
M :	Power of 2 which determines $N = 2^M$
sign :	Sign ± 1 in the exponential in (4.11)

The convention for the array F[i,j] is that j = 0 for the real part of the sample values, and j = 1 for the imaginary parts. The index i runs from 0 to N − 1.

Output parameters:

The values in the array F[i,j] are modified to be the DFT of the values set before entry.

⟩ *4.6.4   Example of use*

The program demonstrates how to take sample values from a function,
store them in an array as required, take the Fourier transform, and then
take the inverse Fourier transform. As the complete output is long only
extracts are given.

```
{ 6-2 fft.pas                          }
{ - FFT simple example - RDH 13/7/89 }

PROGRAM Fftx;

  USES
    Cit_core, Cit_prim, Cit_ctrl, Cit_math;

  CONST
    NMAX = 511;
  TYPE
    farray = ARRAY [0..NMAX, 0..1] OF Citreal;
  VAR
    functrans : farray;

  VAR
    power2, npts : integer;
    trange, frange : Citreal;
    ch : char;

  FUNCTION x (t : Citreal) : Citreal;
    VAR
      tol : Citreal;
  BEGIN
    tol:=EPSILON * trange;
    IF t < (- 1.0 - tol) THEN
      x := 0.0
    ELSE
      IF t < (- 1.0 + tol) THEN
        x := 0.5
      ELSE
```

```
        IF t < (1.0 - tol) THEN
          x := 1.0
        ELSE
          IF t < (1.0 + tol) THEN
            x := 0.5
          ELSE
            x := 0.0
  END;

  { sets up complex values in
    array FuncTrans[index, 0=real/1=imag] }
  PROCEDURE setf;
    VAR
      n1, i : integer;
      dt    : Citreal;
  BEGIN
    n1 := npts DIV 2;
    dt := trange / Float (npts);
    FOR i := n1 + 1 TO npts - 1 DO BEGIN
      functrans [i, 0] :=
        x (dt * Float (i - npts));
      functrans [i, 1] := 0.0
    END;
    FOR i := 0 TO n1 DO BEGIN
      functrans [i, 0] := x (dt * Float (i));
      functrans [i, 1] := 0.0
    END;
  END { setF };

  PROCEDURE settrans (k, sign : Citreal);
    VAR
      i : integer;
  BEGIN
    FOR i := 0 TO npts - 1 DO BEGIN
      functrans [i, 0] := k * functrans [i, 0];
      functrans [i, 1] := k * functrans [i, 1]
    END;
    Fft (@functrans, power2, sign);
  END;
```

```
PROCEDURE display_data;
BEGIN
  npts := Round (Raise (2, power2));
  Writeln ('Power of 2: M =', power2:3,
    ' so N =', npts:6);
  frange := npts / trange;
  Writeln ('T =', trange:6:2,
    ' so freq range F=', frange:6:2);
END;


{of display_data}
PROCEDURE print_values;
  VAR
    i : integer;
BEGIN
  Writeln (' i     real        imag');
  FOR i := 0 TO npts - 1 DO
    Writeln (i:3, '  ', functrans [i, 0]:10,
      '  ', functrans [i, 1]:10);
END;


{==================== MAIN ========================}
BEGIN
  power2 := 4;
  trange := 4.0;
  Text_mode;
  display_data;
  setf;
  print_values;
  {unit scale, sign=-1 to find XT}
  settrans (1, - 1);


  Pause;
  Text_mode;
  {resets screen colours}
  print_values {scale=1/N, sign=+1 to find XTT};
  settrans (1, 1);
```

```
    Pause;
    Text_mode;
    print_values


END.


DISCRETE FOURIER TRANSFORM
Power of 2: M =  4 so N =    16
T =  4.00 so freq range F=  4.00
i     real         imag
0    1.000E+00    0.000E+00
1    1.000E+00    0.000E+00
2    1.000E+00    0.000E+00
3    1.000E+00    0.000E+00
4    5.000E-01    0.000E+00
5    0.000E+00    0.000E+00
6    0.000E+00    0.000E+00
...

i     real         imag
0    8.000E+00    0.000E+00
1    5.027E+00   -1.819E-12
2    0.000E+00    0.000E+00
3   -1.497E+00   -3.638E-12
4    0.000E+00    0.000E+00
5    6.682E-01   -6.821E-12
6    0.000E+00    0.000E+00
...

i     real         imag
0    1.000E+00    0.000E+00
1    1.000E+00   -4.547E-13
2    1.000E+00    2.501E-12
3    1.000E+00    4.547E-12
4    5.000E-01    0.000E+00
5    6.366E-12    2.274E-12
6    7.276E-12   -1.819E-12
...
```

The inverse transform should reproduce the original sample values, but we do not get the expected zero values due to the limits of machine accuracy. However the deviations are of the same relative order as the machine rounding error, as might be hoped. The results shown above were obtained using arithmetic with a 'machine epsilon' of approximately **2E-12**.

⟩ Chapter 5

⟩ Mathematical Routines II

⟩ 5.1 Solution of linear equations

⟩ 5.1.1 Matrices and Gaussian elimination

Linear equations occur in all kinds of mathematical problems. They are sets of equations like

$$
\begin{array}{ccccccc}
x & + & 2y & + & 3z & = & 14 \\
2x & + & 3y & + & 4z & = & 20 \\
x & + & y & + & 2z & = & 9
\end{array}
$$

The quantities x, y and z are called the *unknowns*, and they must be given values that satisfy all the equations at the same time. In this example, the solution is $x = 1$, $y = 2$, $z = 3$.

There should be at least two variables, but there is no upper limit and some industrial applications involve thousands of variables. Our routines are not intended for that sort of problem!

Linear equations can be neatly set out in matrix form. The above example then becomes

$$
\begin{pmatrix} 1 & 2 & 3 \\ 2 & 3 & 4 \\ 1 & 1 & 2 \end{pmatrix} \times \begin{pmatrix} x \\ y \\ z \end{pmatrix} = \begin{pmatrix} 14 \\ 20 \\ 9 \end{pmatrix}
$$

Symbolically

$$
\mathbf{Ax} = \mathbf{y} \quad \text{or} \quad \sum_j A_{ij} x_j = y_j
$$

A is the matrix, **x** is the vector of unknowns, **y** is the *right-hand-side* vector, and a_{ij}, x_j, y_j denote the elements of each of these arrays. If n is the number of unknowns (and therefore also the number of equations), then i and j run from 1 to n, and n is called the *order of the matrix*.

For sets of equations up to quite large orders one method of solution is known as *Gaussian elimination*. The first equation is used to give the first unknown in terms of the rest. This expression is then substituted into the other equations and the algebra tidied up leaving a set of $n-1$ equations in $n-1$ unknowns. The first variable has thus been eliminated. This process can be repeated until there is only one unknown in one remaining equation, and it is easily solved. There are other techniques for various special cases.

A thorough discussion of numerical methods for solving linear equations is beyond the scope of this book. The section *Further reading* in the *Introduction* suggests suitable references.

The Toolkit provides two sets of procedures. The routine ELIM carries out Gaussian elimination with the refinement known as *pivoting*. It improves the accuracy and deals with some cases which would otherwise fail. ELIM should generally be satisfactory in simple cases, but it will not warn you if the matrix is nearly *singular*, i.e. if the equations are close to not having a solution, something which can happen frequently.

This problem can be overcome with a more powerful approach using the two routines DECOMP and SOLVE. DECOMP is called first. It begins by using Gaussian elimination, but only on the matrix and not on the right-hand-side vector **y**. After the necessary operations have been performed, the matrix is said to have been *decomposed*. DECOMP also estimates a number known as the CONDition of the matrix. This shows by how much errors (e.g. rounding-up errors) are magnified in the solution, and so indicates how well the problem can be solved. A value over 1E8 means that the matrix is singular or virtually so. However if COND is not too large, SOLVE can be used to find the solution. The advantage of this two-stage approach is that the decomposition need only be done once, and it can then easily be applied to solve the equation with different right-hand sides.

The solution may also be expressed mathematically using a matrix related to **A** known as the *inverse matrix* of **A**, often written $\mathbf{A}^{-1}$. Symbolically $\mathbf{x} = \mathbf{A}^{-1}\mathbf{y}$.

In most applications it is the solution of a set of linear equations that is wanted, but sometimes the inverse matrix itself is required, and this is not found by the Gaussian elimination process described above. Another procedure **INVERT** is provided for this purpose. It could also be used to solve linear equations but that would be slower than the elimination method.

) 5.1.2   *Working example*

Filename: **1-1MATRI**
To run the working example see *Guide to running the software*.

1-1MATRI is a comprehensive demonstration of all the routines mentioned above: **ELIM, DECOMP** and **SOLVE**, and **INVERT**. It is driven at the top level by means of a menu bar which reads:

Options: Order read matrix RHS save solution invert quit

Order prompts for a new order of matrix (up to $n = 19$ is allowed). The option **read** allows a matrix previously saved on a file to be read back, and the order will also be reset. Option **matrix** displays the matrix and allows its values to be edited. If it is too large to fit on the screen all at once, the standard location commands can be used to move around (a summary of these is displayed at the foot of the screen). Option **RHS** permits editing of the right-hand-side vector **y**. Note that the right-hand-side vector may also be reset by the **read** option, if the file contained data for the right-hand side. Option **save** writes the order, matrix, and right-hand-side vector to file. Option **solution** brings up a menu allowing choice of method, after which the solution, determinant of **A**, and condition number are displayed. Option **invert** calculates A^{-1} and displays it.

Note that if the inverse is displayed when option **save** is invoked, then it is the inverse matrix which is written to file. The un-inverted matrix can be recalled by using the **matrix** option. An interesting test of the **INVERT** procedure can be made by inverting the original matrix, saving it to a file, then reading it back so it becomes the new matrix, and then selecting the **invert** option again. This forces the program to invert the inverse matrix, whereas selecting the inverse option whilst the inverse is displayed on screen merely restores the original data. If the computer arithmetic was perfect, the original matrix would be recovered. To the extent that it is not, you can see the effect of rounding errors.

A full discussion of the numerical analysis of matrix methods is beyond the scope of this book: consult *A Simple Introduction to Numerical Analysis* (mentioned in *Further reading* in the *Introduction*).

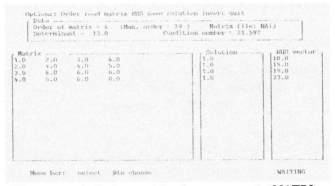

Figure 5.1 A display resulting from program 1-1MATRI

⟩ 5.1.3 Routines ELIM, DECOMP and SOLVE, INVERT

⟩ Procedure ELIM

Routine name: **ELIM**
Kept in unit: **CIT_MATR**
Purpose: To solve linear equations using
                  Gaussian elimination with pivoting

The unit name **CIT_MATR** must be declared in the **Uses** ... statement at the head of your program.

ELIM attempts to solve equations of the form

$$\mathbf{Ax} = \mathbf{y}, \text{ equivalent to } \sum_j a_{ij} x_j = y_i$$

ELIM also finds the determinant of **A**. Solutions of the equations exist and will be found by **ELIM** unless **A** is *singular*, in which case the determinant will be zero and the routine will exit as soon as this condition is detected. Matrices that are nearly singular may not be detected: the determinant will usually be large in these situations and it will be better to use **DECOMP** and **SOLVE**.

ELIM automatically allocates the workspace which it needs from the heap and does not alter the values stored in the arrays used to store **A** and **y**.

An exit on zero determinant is *not* treated as an error exit, so **Errorflag** is not set in this case. An error return from **ELIM** can only occur if there is insufficient heap space to make local copies of the arrays A and Y.

Signature:

```
procedure ELIM(n,n0: integer:
        Aptr,Xptr,Yptr: Arrayptr;
                VAR det: Citreal);
```

Input parameters:

Call:	ELIM(n,n0,@A,@X,@Y,det)
n:	Order of the matrix
n0:	Row dimension of the array A
@A:	Pointer to array used for matrix
@X:	Pointer to array to hold solution
@Y:	Pointer to array used for right-hand-side **y**

Output parameters:

@X:	Pointer to array holding solution
det:	Variable set to value of determinant
	det $= 0$ if matrix singularity was detected

Note: the contents of arrays A and Y are *not* altered.

Before calling the routine the user must set up the matrix and right-hand-side vector in suitable arrays, and have another array ready to hold the solution vector. In the parameter descriptions above the names A, Y, and X have been used for these purposes. Using these names, a_{ij} would be stored in A[i,j], y_j in Y[j], *etc.* Note that the contents of arrays A and Y are *not* changed. An example of a suitable declaration is:

```
const
    m0=15;
    n0=10;
var
    A: array[1..m0,1..n0] of Citreal;
    X,Y: array[1..m0] of Citreal;
```

Since linear equations require a square matrix if there is to be a solution (number of equations = number of unknowns), it would be usual for n0 to equal m0, their value being the maximum order of the matrix that could be stored in the array A as dimensioned. However the routine does not *require* that this is the case: in the above example the maximum order would be 10 and the last 5 rows of A are redundant (for current purposes). Note that ELIM does not need to know m0, although it does assume that m0 > n0. If this is not the case then ELIM will alter memory locations outside those allocated for A and there will be unpredictable consequences.

) *Procedures DECOMP and SOLVE*

Routine name: DECOMP, SOLVE
Kept in unit: CIT_MATR
Purpose: Solution of linear equations with estimation of condition number

The unit name CIT_MATR must be declared in the Uses ... statement at the head of your program.

Procedures DECOMP and SOLVE have been adapted from *Computer Methods for Mathematical Computations* by Forsythe, Malcolm and Moler (Prentice-Hall, 1977).

The form of the equations is the same as for ELIM:

$$\mathbf{Ax} = \mathbf{y}, \text{ equivalent to } \sum_j a_{ij} x_j = y_i$$

The procedures should be used in two stages. First call DECOMP to decompose the matrix, then if the condition number cond is not too large SOLVE can be used to find the solution. cond shows by how much errors (e.g. rounding-up errors) are magnified in the solution, and a very large value (1E8, for example) means that the matrix is singular or virtually so. Exact singularity is indicated by a zero determinant. The matrix need only be decomposed once, and then it can easily be applied to solve the equations with various right-hand sides.

DECOMP *does not* allocate its own workspace: the array (A[i,j] for example) in which values of A are passed is used for workspace. On exit this array stores the decomposed matrix. Similarly SOLVE modifies the array in which the values of y are passed so that they hold the solution on exit.

An exit from DECOMP on zero determinant is *not* treated as an error exit, so Errorflag is not set in this case. However an attempt to call SOLVE using a singular decomposed matrix will result in an error return, but this can only occur if the calling program has not checked that det is non-zero.

Signatures:

```
procedure DECOMP(n,n0: integer:
                      Aptr: Arrayptr;
           VAR det, cond: Citreal);

procedure  SOLVE(n,n0: integer:
           Aptr,Xptr: Arrayptr);
```

Input parameters:

Call:	DECOMP(n,n0,@A,det,cond)
n:	Order of the matrix
n0:	Row dimension of the array A
@A:	Pointer to array holding matrix
Call:	SOLVE(n,n0,@A,@X)
n:	Order of the matrix
n0:	Row dimension of the array A
@A:	Pointer to array holding decomposed matrix
@X:	Pointer to array holding right-hand-side vector

Output parameters for DECOMP:

@A:	Pointer to array holding decomposed matrix
det:	Variable set to value of determinant
	det = 0 if matrix singularity was detected
cond:	Variable set to value of condition

Output parameters for SOLVE:

@X:	Pointer to array holding solution

Before calling DECOMP the user must set up the matrix in a suitable array which will then be modified to hold the decomposed matrix. Similarly before calling SOLVE another array must be set to the right-hand-side vector, and this array will be modified to hold the solution. The notes on the setting and dimensioning of arrays given in the description of ELIM apply to DECOMP and SOLVE too, except that only one vector array is strictly needed; i.e. in the example used for ELIM, X could be omitted and Y used for the right-hand-side vector *and* the solution. However it is better programming practice to retain the right-hand-side values and copy them into the solution array X before invoking SOLVE. For example:

```
X := Y;
SOLVE(n,n0,@A,@X);
```

⟩ *Procedure INVERT*

Routine name: **INVERT**
Kept in unit: **CIT_MATR**
Purpose: Invert a matrix

The unit name **CIT_MATR** must be declared in the **Uses** ... statement at the head of your program.

The procedure inverts a square matrix and calculates its determinant and condition number. If the matrix is singular and therefore has no inverse then the procedure exits with **det** set to zero as soon as this condition is detected.

The procedure operates directly on the array storing the matrix. The result (i.e. the inverted matrix) therefore overwrites the original matrix.

Signature:

```
procedure INVERT(n,n0: integer:
                 Aptr: Arrayptr;
             VAR det,cond: Citreal);
```

Input parameters:

Call:	INVERT(n,n0,@A,det,cond)
n:	Order of the matrix
n0:	Row dimension of the array **A**
@A:	Pointer to array used for matrix

Output parameters:

@A:	Pointer to array holding inverted matrix
det:	Variable set to value of determinant
	det $= 0$ if matrix singularity was detected
cond:	Variable set to value of condition

⟩ *5.1.4   Examples of use*

⟩ *Simple equation solved with ELIM*

The program copies the matrix coefficients from preset constants and then solves for the unknowns.

```
{ 1-2 elim.pas                            }
{ -  Simple use of ELIM - RDH 20/3/89 }

PROGRAM Elimex;

  USES
    Cit_core, Cit_matr;

  { Set up arrays and data }
  CONST
    NMAX = 5;
    NPRESET = 3;
  TYPE
    matrix_array = ARRAY [1..NMAX,
      1..NMAX] OF Citreal;
    vector = ARRAY [1..NMAX] OF Citreal;
    preset_vec = ARRAY [1..NPRESET] OF Citreal;
    preset_array = ARRAY [1..NPRESET,
      1..NPRESET] OF Citreal;
  CONST
    PRESET_VALUES:preset_array = ((1, 4, 9), (9,
      16, 25), (25, 36, 49));
    PRESET_RHS:preset_vec = (14, 50, 110);

  VAR
    matrixa : matrix_array;
    rhs, xvec : vector;
    aptr, rhsptr, xptr : Arrayptr;

    nn  {Order} , i, j : integer;
    det : Citreal;

BEGIN
  {====== main ======}
  nn := NPRESET;
  FOR i := 1 TO nn DO BEGIN
    FOR j := 1 TO nn DO
      matrixa [i, j] := PRESET_VALUES [i, j];
    rhs [i] := PRESET_RHS [i]
  END;
```

```
{set the array pointers}
aptr := @matrixa;
rhsptr := @rhs;
xptr := @xvec;

Elim (nn, NMAX, aptr, rhsptr, xptr, det);

IF det <> 0.0 THEN BEGIN
  Writeln ('Solution:');
  FOR i := 1 TO nn DO
    Writeln ('  X[', i:1, '] = ',
      xvec [i]:12:10);
  Writeln ('Determinant = ', det:14:8)
END
ELSE
  Writeln ('Matrix is singular.')

END.

Solution:
  X[1] = 1.0000000000
  X[2] = 1.0000000000
  X[3] = 1.0000000000
Determinant =   -64.00000000
```

) *Simple use of DECOMP and SOLVE*

```
{ 1-3 decomp.pas                                     }
{ - Simple use of DECOMP and SOLVE - RDH 15/2/89 }


PROGRAM Decompex;

  USES
    Cit_core, Cit_matr;
```

```
{ Set up arrays and data }
CONST
  NMAX = 5;
  NPRESET = 3;
TYPE
  matrix_array = ARRAY [1..NMAX,
    1..NMAX] OF Citreal;
  vector = ARRAY [1..NMAX] OF Citreal;
  preset_vec = ARRAY [1..NPRESET] OF Citreal;
  preset_array = ARRAY [1..NPRESET,
    1..NPRESET] OF Citreal;
CONST
  PRESET_VALUES:preset_array = ((1, 2, 3), (2,
    3, 4), (1, 1, 2));
  PRESET_RHS:preset_vec = (14, 20, 9);


VAR
  matrixa : matrix_array;
  xvec : vector;
  aptr, xptr : Arrayptr;


  nn  {Order} , i, j : integer;
  det, cond : Citreal;

BEGIN

  {====== main ======}
  nn := NPRESET;
  FOR i := 1 TO nn DO BEGIN
    FOR j := 1 TO nn DO
      matrixa [i, j] := PRESET_VALUES [i, j];
    xvec [i] := PRESET_RHS [i]
  END;

  {set the array pointers}
  aptr := @matrixa;
  xptr := @xvec;
```

```
Decomp (nn, NMAX, aptr, det, cond);
Writeln ('Determinant = ', det:14:8);
Writeln ('Condition   = ', cond:14:8);


IF cond < 1E8 THEN BEGIN
  Solve (nn, NMAX, aptr, xptr);
  Writeln ('Solution:');
  FOR i := 1 TO nn DO
    Writeln ('  X[', i:1, '] = ',
      xvec [i]:12:10);
END
ELSE
  Writeln ('Matrix is ill-conditioned.')

END.

Determinant =   -64.00000000
Condition   =    21.31578947
Solution:
  X[1] = 1.0000000000
  X[2] = 2.0000000000
  X[3] = 3.0000000000
```

⟩ *An example of the use of INVERT*

```
{ 1-4 invert.pas                        }
{ -  Simple use of INVERT - RDH 15/2/89 }


PROGRAM Invertex;

  USES
    Cit_core, Cit_form, Cit_matr;
```

```
{ Set up arrays and data }
CONST
  NMAX = 5;
  NPRESET = 4;
TYPE
  matrix_array = ARRAY [1..NMAX,
    1..NMAX] OF Citreal;
  preset_array = ARRAY [1..NPRESET,
    1..NPRESET] OF Citreal;
CONST
  PRESET_VALUES:preset_array = ((1, 2, 3, 4),
    (3, 4, 5, 6), (5, 5, 6, 7), (8, 3, 7, 4));
VAR
  matrixa, matrixp : matrix_array;
  aptr, pptr : Arrayptr;

  nn  {Order} , i, j, k : integer;
  det, cond, p : Citreal;

PROCEDURE printmatrix (VAR a : matrix_array);
  VAR
    i, j : integer;
    str : Linestring;
  BEGIN
    FOR i := 1 TO nn DO BEGIN
      FOR j := 1 TO nn DO BEGIN
        Format_real (a [i, j], 12, ' ', str);
        Write (str + '     ')
      END;
      Writeln
    END
  END;

BEGIN

  {====== main ======}
  nn := NPRESET;
  FOR i := 1 TO nn DO
    FOR j := 1 TO nn DO
      matrixa [i, j] := PRESET_VALUES [i, j];
```

```
{set the array pointers}
aptr := @matrixa;
pptr := @matrixp;

Writeln ('Original matrix:');
printmatrix (matrixa);

Invert (nn, NMAX, aptr, det, cond);
Writeln;
Writeln ('Determinant = ', det:14:8);
Writeln ('Condition   = ', cond:14:8);

IF cond > 1E8 THEN
  Writeln ('Matrix is ill-conditioned.')
ELSE BEGIN
  Writeln;
  Writeln ('Inverse matrix:');
  printmatrix (matrixa);
  Writeln;
  Writeln ('Check A*INV(A):');
  FOR i := 1 TO nn DO
    FOR j := 1 TO nn DO BEGIN
      p := 0.0;
      FOR k := 1 TO nn DO
        p := p + PRESET_VALUES [i,
          k] * matrixa [k, j];
      matrixp [i, j] := p
    END;
  printmatrix (matrixp)
END

END.
```

```
Original matrix:
1.0            2.0            3.0            4.0
3.0            4.0            5.0            6.0
5.0            5.0            6.0            7.0
8.0            3.0            7.0            4.0
Determinant =  14.000000
Condition   = 121.571429
```

```
Inverse matrix:
   0.5               -1.5              1.0              1.36424E-12
  -1.928571428        2.214285714     -0.714285714     -0.142857143
  -0.642857143        2.071428571     -1.571428571      0.285714286
   1.571428571       -2.285714285      1.285714285     -0.142857143

Check A*INV(A):
   1.0                0.0             -1.45519E-11     -9.09495E-13
  -1.45519E-11        1.0              0.0             -1.81899E-12
   0.0                2.91038E-11      1.0             -9.09495E-13
   0.0                1.45519E-11      0.0              1.0
```

The final calculation is done as a check; the matrix product of **A** and its inverse should be the *unit matrix*, that is the matrix with 1 on its diagonal and 0 everywhere else. This is what is obtained, apart from a rounding error; the largest error being **3.72529E-9** at the **[3,1]** and **[4,2]** positions.

⟩ 5.2   Matrix eigenvalues

⟩ 5.2.1   *Eigenvectors*

Suppose that **A** is an n by n matrix, **x** and **y** are vectors, and that they are related by

$$\mathbf{A}\mathbf{x} = \mathbf{y}$$

If **A** has elements a_{ij} and **x** and **y** have components x_i and y_i, this relationship can be expressed in terms of summations over the subscripts:

$$y_i = \sum_{j=1}^{n} a_{ij} x_j$$

If an **x** can be found so that **x** and **y** are parallel, there will be a constant λ such that **y** $= \lambda$**x**, and then λ is an *eigenvalue* of **A**, and **x** is an *eigenvector*. As a matrix equation can be multiplied throughout by a constant, it will also be true that any vector parallel to **x** is also an eigenvector. Thus only the direction of the vector is important not the length, and we are free to choose a sensible standard length for

x. From now on all eigenvectors will be normalised, that is made unit vectors. Any non-zero vector can be normalised simply by dividing all its components by its original length:

$$u_i = \frac{x_i}{\left[\sum_{j=1}^{n} x_j^2\right]^{\frac{1}{2}}}$$

The problem is to find the eigenvectors and eigenvalues of matrix **A**. This chapter will describe three methods. Two simple ones are described by examples in this section, and then a more powerful method is given for the special case when **A** is a real symmetric matrix. (*Symmetric* means that $a_{ij} = a_{ji}$. In most cases that arise in practice where eigenvalues are required, the matrix concerned is symmetric.) In general, eigenvectors may not exist, but it is known that they cannot number more than n. For a symmetric matrix, there are always n eigenvectors.

⟩ *The power method of finding eigenvectors*

The first method is sometimes known as the *power method*. It finds the eigenvector corresponding to the eigenvalue with the largest absolute value; it assumes both that one exists and that it is unique. It is based on the rule that if any vector **x** is repeatedly multiplied by **A**, its direction gradually swings around coming closer and closer to the eigenvector. That is, one repeatedly applies the rule:

$$\mathbf{x}^{(2)} := \mathbf{A}\mathbf{x}^{(1)}, \ \mathbf{x}^{(1)} := \frac{\mathbf{x}^{(2)}}{|\mathbf{x}^{(2)}|}$$

until the directions of $\mathbf{x}^{(1)}$ and $\mathbf{x}^{(2)}$ are virtually the same.

In the programmed example that follows, X1[j] is the initial vector, which is multiplied by A[i,j] to give the new vector X2[i]. This is normalised and made the new initial vector. This process is continually repeated until X1[i] and X2[i] are sufficiently close to each other (after normalisation). X2[i] is then taken as the eigenvector and the normalisation constant gives the absolute value of the eigenvalue (the sign of the eigenvalue may be found by inspecting the last two iterates; if the vector changes sign, the eigenvalue is negative). This example gives the essential parts of the library routine POWER described later.

```
{ 2-1 power.pas                                        }
{ -  Power method for Eigenvalues - RDH 15/2/89 }
```

```
PROGRAM Powerex;

  USES
    Cit_core, Cit_form, Cit_matr;

  { Set up arrays and data }
  CONST
    NMAX = 3;
  TYPE
    matrix_array = ARRAY [1..NMAX,
      1..NMAX] OF Citreal;
    vector = ARRAY [1..NMAX] OF Citreal;
  CONST
    MATRIXA:matrix_array = ((1, 2, 3), (1, 4, 9),
      (1, 8, 27));
    INITX:vector    = (1, 1, 1);

  VAR
    x1, x2 : vector;
    i, j : integer;
    lambda, modx, deltax, sum : Citreal;

  PROCEDURE printvector (pre : Linestring;
                         VAR x : vector);
    VAR
      i : integer;
      str : Linestring;
  BEGIN
    Write (pre);
    FOR i := 1 TO NMAX DO BEGIN
      Format_real (x [i], 12, ' ', str);
      Write (str + '    ')
    END;
    Writeln
  END;

BEGIN
  {====== main ======}
  Writeln ('Power Method:');

  { Initialise X2 }
  x2 := INITX;
```

```
{Iteration loop}
REPEAT
  { copy vector X2 to X1 }
  x1 := x2;

  { matrix multiply: X2:=A*X1 }
  FOR i := 1 TO NMAX DO BEGIN
    sum := 0.0;
    FOR j := 1 TO NMAX DO
      sum := sum + MATRIXA [i, j] * x1 [j];
    x2 [i] := sum
  END;

  { find length of X2 }
  modx := 0.0;
  FOR i := 1 TO NMAX DO
    modx := modx + Sqr (x2 [i]);
  modx := Sqrt (modx);

  { normalise X2 }
  FOR i := 1 TO NMAX DO
    x2 [i] := x2 [i] / modx;

  { find how much X1 has changed }
  deltax := 0.0;
  FOR i := 1 TO NMAX DO
    deltax :=
      deltax + Abs (Abs (x1 [i]) - Abs (x2 [i]))
    ;

UNTIL deltax < Sqrt (EPSILON);

Writeln ('Last 2 iterates were:');
printvector ('X1 = ', x1);
printvector ('X2 = ', x2);
Writeln ('Absolute value of eigenvalue = ',
  modx:10:6)

END.
```

```
Power Method:
Last 2 iterates were:
X1 =    0.119842326 0.329512716   0.936514382
X2 =    0.119842326 0.329512716   0.936514382

Absolute value of eigenvalue = 29.942767
```

) *The modified power method*

The second method is a modification of the first and allows any of the eigenvalues to be found. It uses the result that if I is the unit matrix and λ_0 is any number, then the matrix $[A - \lambda_0 I]^{-1}$ has eigenvalues $\mu = \frac{1}{(\lambda - \lambda_0)}$. The largest value of μ corresponds to the eigenvector for which $|\lambda - \lambda_0|$ is least. By varying the choice of λ_0, it is possible to make the power method, when applied to the modified matrix, converge to any of the eigenvectors. Once the eigenvalue μ of the modified matrix has been found, λ is given by $\lambda = \lambda_0 + \frac{1}{\mu}$.

It is not necessary to compute the inverse matrix, instead, at each stage of the iteration, one solves

$$[A - \lambda_0 I]x^{(2)} = x^{(1)}$$

The routines DECOMP and SOLVE (see Section 5.3) are ideal for this purpose, as the matrix $[A - \lambda_0 I]$ needs to be decomposed only once at the beginning of the iteration. This example also incorporates an automatic test for the sign of the eigenvalue μ, which is necessary for the correct value of λ. In essence this example is the same as used in the library routine MODPOWER.

Note that the variable L is used for λ_0.

```
{ 2-2 modpower.pas                              }
{ - Modified Power method - RDH 13/7/89 }

PROGRAM Modpowerex;

    USES
        Cit_core, Cit_form, Cit_matr;
```

```
{ Set up arrays and data }
CONST
  NMAX = 5;
  NPRESET = 3;
TYPE
  matrix_array = ARRAY [1..NMAX,
    1..NMAX] OF Citreal;
  vector = ARRAY [1..NMAX] OF Citreal;
  preset_vec = ARRAY [1..NPRESET] OF Citreal;
  preset_array = ARRAY [1..NPRESET,
    1..NPRESET] OF Citreal;
CONST
  PRESET_VALUES:preset_array = ((1, 2, 3), (1,
    4, 9), (1, 8, 27));
  INITX:preset_vec = (1, 1, 1);


VAR
  matrixa : matrix_array;
  x1, x2 : vector;
  aptr, xptr : Arrayptr;


  nn  {Order} , i, j : integer;
  det, cond : Citreal;
  l, modx, deltax, sign : Citreal;


PROCEDURE printvector (pre : Linestring;
                          VAR x : vector);
  VAR
    i : integer;
    str : Linestring;
BEGIN
  Write (pre);
  FOR i := 1 TO nn DO BEGIN
    Format_real (x [i], 12, ' ', str);
    Write (str + '    ')
  END;
  Writeln
END;
```

```
BEGIN

  {====== main ======}
  Writeln ('Modified Power Method:');
  nn := NPRESET;

  { Input }
  Write ('Input value for L: ');
  Readln (l);

  { Initialise matrix and X2 }
  FOR i := 1 TO nn DO BEGIN
    FOR j := 1 TO nn DO BEGIN
      matrixa [i, j] := PRESET_VALUES [i, j];
      IF i = j THEN
        matrixa [i, i] := matrixa [i, i] - 1
    END;
    x2 [i] := INITX [i]
  END;

  {set the array pointers}
  aptr := @matrixa;
  xptr := @x2;

  { decompose matrix }
  Decomp (nn, NMAX, aptr, det, cond);
  Writeln ('Determinant = ', det:14:8);
  Writeln ('Condition   = ', cond:14:8);

  IF cond > 1E8 THEN BEGIN
    Writeln ('Matrix is ill-conditioned.');
    Exit
  END;

  {Iteration loop}
  REPEAT
    { copy vector X2 to X1 }
    x1 := x2;

    { solve (A-LI)*X2=X1 }
    Solve (nn, NMAX, aptr, xptr);
```

```
{ find length of X2 }
modx := 0.0;
FOR i := 1 TO nn DO
  modx := modx + Sqr (x2 [i]);
modx := Sqrt (modx);

{ normalise X2 }
FOR i := 1 TO nn DO
  x2 [i] := x2 [i] / modx;

{ find how much X1 has changed }
deltax := 0.0;
FOR i := 1 TO nn DO
  deltax :=
    deltax + Abs (Abs (x1 [i]) - Abs (x2 [i]))
  ;

UNTIL deltax < Sqrt (EPSILON);

Writeln ('Last 2 iterates were:');
printvector ('X1 = ', x1);
printvector ('X2 = ', x2);

{ find sign of modified eigenvalue      }
{ by comparing first non-zero elements }
{ of X1 and X2.                         }
i := 0;
REPEAT
  i := i + 1;
  sign := Sgn (x1 [i] * x2 [i])
UNTIL Abs (x2 [i]) > 0.1;

Writeln ('Eigenvalue = ',
  (1 + sign / modx):10:6)
END.
```

```
Modified Power Method:
Input value for L: 29.9
Determinant =    35.62100000
Condition   = 1113.26927097
Last 2 iterates were:
X1 =   0.119842326 0.329512716  0.936514382
X2 =   0.119842326 0.329512716  0.936514382
Eigenvalue = 29.942767
```

```
Modified Power Method:
Input value for L: 0
Determinant =    12.00000000
Condition   =    84.50000000
Last 2 iterates were:
X1 =   0.848423186 -0.515028274  0.122163723
X2 =   0.848423186 -0.515028274  0.122163723
Eigenvalue =  0.217884

Modified Power Method:
Input value for L: 1.8
Determinant =    -1.75200000
Condition   =   598.95891057
Last 2 iterates were:
X1=   0.716252231   0.656311363  -0.237145813
X2=   0.716252231   0.656311363  -0.237145813
Eigenvalue = 1.839349
```

The result shows that the test matrix **A** had three real eigenvalues and they have all been found, together with their eigenvectors. The method is simple, but does suffer from some uncertain convergence behaviour, particularly when two eigenvalues are close together. It does not help in making the choice of λ_0 either. Note that the method fails if λ_0 is chosen exactly equal to an eigenvalue.

) *Jacobi's method*

The third method is Jacobi's method which applies to real symmetric matrices, that is those that have real elements a_{ij} satisfying $a_{ij} = a_{ji}$. The coordinate axes are rotated according to a certain formula, and **A** is modified accordingly until it is a diagonal matrix. The eigenvectors are the new coordinate axes and the eigenvalues are the diagonal elements of the modified **A**. Thus Jacobi's method finds all the eigenvalues and eigenvectors at one application. If **A** is not symmetric then the routine first takes the symmetric part using

$$a_{ij} := \frac{1}{2}(a_{ij} + a_{ji})$$

The routine JACOBI implements this method. A full description of Jacobi's method is beyond the scope of this book.

⟩ *5.2.2*   *Working example*

Filename: `2-3EIGEN.PAS`
To run the working example see *Guide to running the software.*

2-3EIGEN is a comprehensive demonstration of all the methods mentioned above: **POWER**, **MODPOWER** and **JACOBI**. It is driven at the top level by means of a menu bar which reads:

Options: **Order read matrix save eigenvalue(s) quit**

Order prompts for a new order of matrix (up to $n = 19$ is allowed). The option **read** allows a matrix previously saved on a file to be read back, and the order will also be reset. (The file format is compatible with that used by **1-1MATRI**.) Option **matrix** displays the matrix and allows its values to be edited. If it is too large to fit on the screen all at once, the standard location commands can be used to move around (a summary of these is displayed at the foot of the screen). Option **save** writes the order and the matrix to file. Option **eigenvalue(s)** gives a new menu from which a choice of methods can be made. Once the method has been applied, the eigenvalue(s) found are displayed on the screen next to the matrix window, and the eigenvector(s) replace the matrix display. Any previous values are cleared from the screen, except that if the modified power method is used repeatedly then the results accumulate on the screen.

⟩ *5.2.3*   *Routines POWER, MODPOWER, and JACOBI*

⟩ *Procedure POWER*

Routine name: **POWER**
Kept in unit: **CIT_MATR**
Purpose: Find largest eigenvalue of a matrix

The unit name **CIT_MATR** must be declared in the **Uses** ... statement at the head of your program.

The routine **POWER** applies the power method, as described in Section 5.2.1, to find the largest eigenvalue of a matrix.

Signature:

```
procedure POWER(n,n0: integer;
                acc: Citreal;
           Aptr,Xptr: Arrayptr;
           VAR eigval: Citreal);
```

Input parameters:

Call: POWER(n,n0,acc,@A,@X,eigval)
n: Order of the matrix
n0:   Row dimension of the array A
acc: Accuracy required for eigenvector
@A: Pointer to array used for matrix
@X: Pointer to array to holding estimate of eigenvector

Output parameters:

@X: Pointer to array holding eigenvector
eigval: Eigenvalue found

The way in which the matrix elements a_{ij} should be stored in a suitable array (A[i,j] say) is the same as for routine ELIM: see notes in Section 5.1.3. The contents of array A are *not* altered. The array X[i] should be set to a vector in approximately the direction in which the eigenvector is expected. If this is not known, then any reasonable values such as (1,1,1) will do as long as they are not all zero.

If the method fails to converge, the routine exits with **Errorflag** and other error globals appropriately set.

) *Procedure MODPOWER*

Routine name: **MODPOWER**
Kept in unit: **CIT_MATR**
Purpose: Find eigenvalue closest to given value

The unit name **CIT_MATR** must be declared in the **Uses** ... statement at the head of your program.

The routine **MODPOWER** applies the modified power method, as described in Section 5.2.1, to find the eigenvalue of a matrix nearest to an estimate provided by the user.

Signature:

```
procedure MODPOWER(n,n0: integer;
                   acc: Citreal;
            Aptr,Xptr: Arrayptr;
           VAR eigval: Citreal);
```

Input parameters:

Call:	MODPOWER(n,n0,acc,@A,@X,eigval)
n:	Order of the matrix
n0:	Row dimension of the array A
acc:	Accuracy required for eigenvector
@A:	Pointer to array used for matrix
@X:	Pointer to array to hold eigenvector
eigval:	Estimate of eigenvalue

Output parameters:

@X:	Pointer to array holding eigenvector
eigval:	Eigenvalue found

The way in which the matrix elements a_{ij} should be stored in a suitable array (A[i,j] say) is the same as for routine ELIM: see notes in Section 5.1.3. The contents of array A are altered. The array X[i] should be set to a vector in approximately the direction in which the eigenvector is expected. If this is not known, then any reasonable values such as (1,1,1) will do as long as they are not all zero.

If the method fails to converge, the routine exits with **Errorflag** and other error globals appropriately set.

⟩ *Procedure JACOBI*

Routine name:	JACOBI
Kept in unit:	CIT_MATR
Purpose:	Find the eigenvalues and eigenvectors
	of a real symmetric matrix

The unit name **CIT_MATR** must be declared in the **Uses** ... statement at the head of your program.

The routine **JACOBI** applies the Jacobi method, as described in Section 5.2.1, to find all the eigenvalues and eigenvectors of a real symmetric matrix.

Signature:

```
procedure JACOBI(n,n0: integer;
                 vectors: Boolean;
                 Aptr,EVptr: Arrayptr);
```

Input parameters:

Call:	JACOBI(n,n0,true,@A,@Lambdas)
n:	Order of the matrix
n0:	Row dimension of the array A
vectors	If true, find eigenvectors
@A:	Pointer to array used for matrix
@Lambdas:	Pointer to array to hold eigenvalues

Output parameters:

@A:	Pointer to array which now holds eigenvectors if **vectors** was set **true**
@Lambdas:	Pointer to array holding eigenvalues

The way in which the matrix elements a_{ij} should be stored in a suitable array ($A[i,j]$ say) is the same as for routine ELIM: see notes in Section 5.1.3. The contents of array A are altered whether or not **vectors** is set **true**. If eigenvectors have been calculated, they are stored as rows of **A**.

Results are obtained as accurately as machine rounding errors will allow. Eigenvalues are likely to be more accurate than eigenvectors.

⟩ *5.2.4   Examples of use*

⟩ *Simple use of POWER*

The example finds the largest eigenvalue of the same matrix used in the power method example at the beginning of the chapter.

Since the signature of MODPOWER is the same with only minor alterations, the example will also serve for MODPOWER. Besides changing the routine name, the variable Lambda must be set to an estimate of the eigenvalue sought before the routine call.

```
{ 2-4 powrex.pas                             }
{ - use of POWER procedure - RDH 15/2/89 }
```

```
PROGRAM Powerex;

  USES
    Cit_core, Cit_form, Cit_matr;

  { Set up arrays and data }
  CONST
    NMAX = 5;
    NPRESET = 3;
  TYPE
    matrix_array = ARRAY [1..NMAX,
      1..NMAX] OF Citreal;
    vector = ARRAY [1..NMAX] OF Citreal;
    preset_vec = ARRAY [1..NPRESET] OF Citreal;
    preset_array = ARRAY [1..NPRESET,
      1..NPRESET] OF Citreal;
  CONST
    PRESET_VALUES:preset_array = ((1, 2, 3), (1,
      4, 9), (1, 8, 27));
    INITX:preset_vec = (1, 1, 1);

  VAR
    matrixa : matrix_array;
    xvec : vector;
    aptr, xptr : Arrayptr;

    nn  {Order} , i, j : integer;
    lambda, accreqd : Citreal;

  PROCEDURE printvector (VAR x : vector);
    VAR
      i : integer;
      str : Linestring;
  BEGIN
    FOR i := 1 TO nn DO BEGIN
      Format_real (x [i], 12, ' ', str);
      Write (str + '    ')
    END;
    Writeln
  END;
```

```
BEGIN

  {====== main ======}
  Writeln ('Using POWER procedure:');
  nn := NPRESET;

  { Initialise matrix and Xvec }
  FOR i := 1 TO nn DO BEGIN
    FOR j := 1 TO nn DO
      matrixa [i, j] := PRESET_VALUES [i, j];
    xvec [i] := INITX [i]
  END;

  {set the array pointers}
  aptr := @matrixa;
  xptr := @xvec;

  accreqd := Sqrt (EPSILON);
  Power (nn, NMAX, accreqd, aptr, xptr, lambda);

  IF NOT Errorflag THEN BEGIN
    Writeln ('Eigenvector');
    printvector (xvec);
    Writeln ('Eigenvalue = ', lambda:10:6)
  END
  ELSE
    Writeln (Errorstring)

END.

Using POWER procedure:
Eigenvector:
 0.119842326 0.329512716  0.936514382
Eigenvalue = 29.942767
```

) *Simple use of JACOBI*

The example finds the largest eigenvalue of the matrix:

$$\begin{pmatrix} 1 & 2 & 3 & 4 \\ 2 & 3 & 4 & 5 \\ 3 & 4 & 5 & 6 \\ 4 & 5 & 6 & 7 \end{pmatrix}$$

```
{ 2-5 jacobi.pas                              }
{ - use of JACOBI procedure - RDH 15/2/89 }


PROGRAM Jacobiex;

  USES
    Cit_core, Cit_form, Cit_matr;


  { Set up arrays and data }
  CONST
    NMAX = 5;
    NPRESET = 4;
  TYPE
    matrix_array = ARRAY [1..NMAX,
      1..NMAX] OF Citreal;
    vector = ARRAY [1..NMAX] OF Citreal;
    preset_array = ARRAY [1..NPRESET,
      1..NPRESET] OF Citreal;
  CONST
    PRESET_VALUES:preset_array = ((1, 2, 3, 4),
      (2, 3, 4, 5), (3, 4, 5, 6), (4, 5, 6, 7));


  VAR
    matrixa : matrix_array;
    eigenvals : vector;
    aptr, eptr : Arrayptr;
    nn  {Order} , i, j : integer;
```

```
PROCEDURE printvector (VAR x : vector);
  VAR
    i : integer;
    str : Linestring;
BEGIN
  FOR i := 1 TO nn DO BEGIN
    Format_real (x [i], 12, ' ', str);
    Write (str + '     ')
  END;
  Writeln
END;

PROCEDURE printmatrix (VAR a : matrix_array);
  VAR
    i, j : integer;
    str : Linestring;
BEGIN
  FOR i := 1 TO nn DO BEGIN
    FOR j := 1 TO nn DO BEGIN
      Format_real (a [i, j], 12, ' ', str);
      Write (str + '     ')
    END;
    Writeln
  END
END;

BEGIN
  {====== main ======}
  Writeln ('Using JACOBI procedure:');
  nn := NPRESET;

  { Initialise matrix }
  FOR i := 1 TO nn DO
    FOR j := 1 TO nn DO
      matrixa [i, j] := PRESET_VALUES [i, j];

  {set the array pointers}
  aptr := @matrixa;
  eptr := @eigenvals;

  Jacobi (nn, NMAX, True, aptr, eptr);
```

```
IF NOT Errorflag THEN BEGIN
  Writeln ('Eigenvalues:');
  printvector (eigenvals);
  Writeln ('Eigenvectors (in rows):');
  printmatrix (matrixa)
END
ELSE
  Writeln (Errorstring)

END.
```

Using JACOBI procedure:
Eigenvalues:
-1.16515139   3.50200E-20 17.16515139   -7.13951E-19
Eigenvectors (in rows):
 0.775210019   0.342443158 -0.090323703 -0.523090564
-0.364214387   0.790530718 -0.488418276   0.062101945
 0.314721188   0.427472436   0.540223684   0.652974932
 0.409081753 -0.273972962 -0.679299336   0.544190544

Two of the eigenvalues are actually zero, showing that the accuracy was to full machine precision relative to the largest eigenvalue. As remarked before, the eigenvectors are not found so accurately.

⟩ 5.3   Matrix decomposition

⟩ 5.3.1   *Singular value decomposition*

The method described in this chapter is *Singular Value Decomposition* (SVD). Some knowledge of the theory of linear algebra is required for general use of this technique. However, in the next section a particular application, the method of least squares, is described in less specialised terms.

Given an m by n matrix $\mathbf{A}$ (m rows, n columns) the method finds matrices $\mathbf{U}$, $\mathbf{V}$ and $\mathbf{S}$ such that

$$\mathbf{A} = \mathbf{U}\mathbf{S}\mathbf{V}^t \qquad (5.1)$$

Here, $\mathbf{U}$ and $\mathbf{V}$ are m by m and n by n orthogonal matrices, while $\mathbf{S}$ is an m by n diagonal matrix (except on the diagonal, the elements are all zero). The diagonal elements of $\mathbf{S}$, called s_1, s_2, etc., are the singular values of matrix $\mathbf{A}$. $\mathbf{V}^t$ denotes the transpose of matrix $\mathbf{V}$.

The relative sizes of m and n affect the columns of $\mathbf{U}$ and $\mathbf{V}$ which are significant. In examples (5.2) and (5.3) below the matrix elements marked '.' cannot affect the elements of $\mathbf{A}$ because in forming the matrix products they will always be multiplied by zero elements in the middle matrix $\mathbf{S}$.

$$\mathbf{A} = \begin{pmatrix} u_{11} & u_{12} & . & . \\ u_{21} & u_{22} & . & . \\ u_{31} & u_{32} & . & . \\ u_{41} & u_{42} & . & . \end{pmatrix} \begin{pmatrix} s_1 & 0 \\ 0 & s_2 \\ 0 & 0 \\ 0 & 0 \end{pmatrix} \begin{pmatrix} v_{11} & v_{21} \\ v_{12} & v_{22} \end{pmatrix} \tag{5.2}$$

$$\mathbf{A} = \begin{pmatrix} u_{11} & u_{12} \\ u_{21} & u_{22} \end{pmatrix} \begin{pmatrix} s_1 & 0 \\ 0 & s_2 \\ 0 & 0 \\ 0 & 0 \end{pmatrix} \begin{pmatrix} v_{11} & v_{21} & v_{31} & v_{41} \\ v_{12} & v_{22} & v_{32} & v_{42} \\ . & . & . & . \\ . & . & . & . \end{pmatrix} \tag{5.3}$$

Note that it is $\mathbf{V}^t$ which is displayed in (5.3), so that the rows of dots are actually columns of $\mathbf{V}$.

The decomposition of a matrix in this way can be very useful. With it, for instance, the general solution of a set of linear equations could be found, and it also provides an alternative route to calculating eigenvalues and eigenvectors. A good general discussion and references to more detailed ones will be found in *Computer Methods for Mathematical Computations* by Forsythe, Malcolm and Moler (Prentice-Hall, 1977).

⟩ 5.3.2   *Working example of SVD*

Filename: **3-1SVD.PAS**
To run the working example see *Guide to running the software.*

The working example is in the same style as the working examples for the previous two sections and allows you to input or edit a matrix of your choice and then apply the SVD method and view the results. It is driven at the top level by means of a menu bar which reads:

Options: Order read matrix save SVD display quit

Order prompts for new dimensions of the matrix (up to $m, n = 20$ is allowed). The option **read** allows a matrix previously saved on a file to be read back, and the order will also be reset. (The file format is compatible with that used for matrices in earlier sections although they will be square.) Option **matrix** displays the matrix and allows its values to be edited. If it is too large to fit on the screen all at once, the standard location commands can be used to move around (a summary of these is displayed at the foot of the screen). Option **save** writes the dimensions and the matrix to file. Option **SVD** applies the method and displays the singular values. Option **display** gives a new menu allowing any of the matrices **U**, **V** or **A** to be displayed.

A useful exercise is to apply SVD to the matrices used to demonstrate the eigenvalue methods in the previous section. Although this does not use the full generality of the SVD method, it demonstrates the connection.

The next section describes an SVD-based working example for curve fitting.

⟩ *5.3.3   Routine SVD*

Routine name: SVD
Kept in unit: CIT_MATR
Purpose: Singular value decomposition of a matrix

This routine has been adapted from *Computer Methods for Mathematical Computations* by Forsythe, Malcolm and Moler (Prentice-Hall, 1977).

The unit name **CIT_MATR** must be declared in the **Uses** ... statement at the head of your program.

SVD finds matrices **U** and **V** and an array **W**, such that if **A** is an m by n matrix then

$$\mathbf{A} = \mathbf{USV}^t$$

Here, **S** is an m by n diagonal matrix with diagonal elements given by w_i, $i = 1, \ldots, min(m, n)$, and all other elements zero. **U** is an m by m orthogonal matrix, but if $m > n$ the routine only produces the first n columns of **U** (see (5.2) above). **V** is an n by n orthogonal matrix, but if $n > m$ then only the first m rows will have any significance.

It should also be noted that if any of the singular values are zero or very small, then the corresponding column of **U** and row of **V** will not significantly affect **A**.

Signature:

```
procedure SVD(      m,n,n0: integer;
          Aptr,Uptr,Vptr,Wptr: Arrayptr;
                  VAR svdflag: integer);
```

Input parameters:

Call:	SVD(m,n,n0,@A,@U,@V,@W)
m:	Number of rows in matrix **A**
n:	Number of columns in the matrix
n0:	Row dimension of the array **A**
@A:	Pointer to array used for matrix
@U, @V:	Pointers to arrays to hold **U**, **V**
@W:	Pointer to array to hold singular values

Output parameters:

@U, @V:	Pointers to arrays holding **U**, **V**
@W:	Pointer to array holding singular values
svdflag:	Non-zero if any iteration failed

Note: the contents of array **A** are *not* altered. A non-zero value k of svdflag indicates that the iteration for the kth singular value failed.

Although the matrices **A**, **U** and **V** are all mathematically of different dimensions (unless $m = n$), the arrays used to store their values should all be dimensioned the same. For example:

```
CONST mmax=15; nmax=10;
TYPE matrix=array[1..mmax, 1..nmax] of Citreal;
VAR
   A,U,V: matrix;
       W: array[1..mmax] of Citreal;
```

Note that this permits up to **mmax** rows and **nmax** columns although SVD can be called to process smaller matrices. Note that **W[i]** is dimensioned according to the larger of **mmax** and **nmax**. In the above example, **nmax** is the value that should be used for the argument n0.

⟩ 5.3.4   *Example of use*

The program reads an m by n matrix **A** from preset constants, then evaluates and prints its singular value decomposition.

```
{ 3-2 svdex.pas                              }
{ -  use of SVD procedure - RDH 20/2/89 }
PROGRAM Svdex;
  USES
    Cit_core, Cit_form, Cit_matr;
  { Set up arrays and data }
  CONST
    NMAX = 20;
    MPRESET = 5;
    NPRESET = 3;
  TYPE
    matrix_array = ARRAY [1..NMAX,
      1..NMAX] OF Citreal;
    vector = ARRAY [1..NMAX] OF Citreal;
    preset_array = ARRAY [1..MPRESET,
      1..NPRESET] OF Citreal;
  CONST
    PRESET_VALUES:preset_array = ((1, 6, 11), (2,
      7, 12), (3, 8, 13), (4, 9, 14), (5, 10,
      15));
  VAR
    matrixa, matu, matv : matrix_array;
    svals : vector;
    aptr, uptr, vptr, sptr : Arrayptr;
    mm, nn  {Order} , i, j, n, svdflag : integer;
  PROCEDURE printvector (n : integer;
                              x : vector);
    VAR
      i : integer;
      str : Linestring;
  BEGIN
    FOR i := 1 TO n DO BEGIN
      Format_real (x [i], 12, ' ', str);
      Write (str + '      ')
    END;
    Writeln
  END;
```

```
PROCEDURE printmatrix (m, n : integer;
                            VAR a : matrix_array);
  VAR
    i, j : integer;
    str : Linestring;
  BEGIN
  FOR i := 1 TO m DO BEGIN
    FOR j := 1 TO n DO BEGIN
      Format_real (a [i, j], 12, ' ', str);
      Write (str + '    ')
    END;
    Writeln
  END
END;


BEGIN


  {====== main ======}
  Writeln ('Using SVD procedure:');
  nn := NPRESET;
  mm := MPRESET;
  n := Imin (mm, nn);


  FOR i := 1 TO mm DO
    FOR j := 1 TO nn DO
      matrixa [i, j] := PRESET_VALUES [i, j];


  {set all the array pointers}
  aptr := @matrixa;
  uptr := @matu;
  vptr := @matv;
  sptr := @svals;


  Svd (mm, nn, NMAX, aptr, uptr, vptr, sptr,
    svdflag);
```

```
Writeln ('Singular values:');
printvector (n, svals);
Writeln;
Writeln ('Matrix U:');
printmatrix (mm, n, matu);
Writeln;
Writeln ('Matrix V:');
printmatrix (nn, nn, matv);
Writeln;


IF Errorflag THEN
  Writeln (Errorstring)


END.


Singular values:
35.12722333       2.465396696       9.90367E-12

Matrix U:
-0.354557057     -0.688686644       0.020542052
-0.39869637      -0.375554529       0.047770993
-0.442835683     -0.062422415      -0.506198571
-0.486974996      0.250709699       0.786915956
-0.531114309      0.563841814      -0.34903043

Matrix V:
-0.201664911      0.890317133      -0.40824849
-0.516830501      0.257331627       0.816496581
-0.831996092     -0.375653879      -0.40824829
```

Note that only the first three columns of U are printed. The other values in the array may have been used as workspace and their contents are not defined. Note also that since s_3 is very small (= 9.90..E−12), in this case no particular significance should be attached to the third column of U, although the routine has ensured that this column is in fact a unit vector and is orthogonal to the first two columns, as required if U is to be an orthogonal matrix. Similar remarks apply to the third column of V, except that as there is only one column in question its values are determined uniquely by the requirement that V be orthogonal.

) 5.4   Least squares data fitting

) 5.4.1   *Introduction*

The routines needed for the method of least squares have already been described in earlier sections, but the technique of least squares fitting is sufficiently important to warrant a section on its own. Because no new routines are being introduced, the format of this section is different from the others in this chapter.

A general introduction to data fitting has already been given in Section 4.3. In that section, we were concerned with interpolation, which could be briefly described as a technique for finding a smooth function which links a number of points. In this section, we still seek smooth functions but they no longer pass exactly through all the points.

Suppose there are a number of scientific observations subject to a certain amount of scatter, as for instance of the temperature and resistance of a batch of electronic components. If the observations are in the form (x_i, y_i) it might be expected that there is an underlying relationship $y = y(x)$. The problem is then to find the function $y(x)$ giving the best fit to the data.

One of the commonest cases is where the function is a line. This has the general form $y = c_1 + c_2 x$, and it is then necessary to choose c_1 and c_2. This is called *linear regression*. However it is only a special case of a more general technique where the function is a polynomial of degree $(N + 1)$, given by the general form

$$y(x) = c_1 + c_2 x + c_3 x^2 + \dots + c_{N+2} x^{N+1} \tag{5.4}$$

However we need not confine ourselves to powers of x, and the polynomial is in turn a special case of an even more general function built up out of a wide variety of basis functions $B_i(x)$, given by the general form

$$y(x) = c_1 B_1(x) + c_2 B_2(x) + c_3 B_3(x) + \dots = \sum_{j=1}^{N} c_j B_j(x) \tag{5.5}$$

The polynomial case is given by taking $B_j(x) = x^{j-1}$.

If there are M data points, N must be chosen less than M in order to leave some choice for best fitting. Otherwise $y(x)$ could be made to pass through all the observation points (unless two x-values were equal). The method begins with the calculation of the difference R_i between each observation y and the function value $y(x_i)$:

$$R_i = y_i - \sum_{j=1}^{N} c_j B_j(x_i) \tag{5.6}$$

Then comes the evaluation of the sum of the squares of these differences

$$\sum_{i=1}^{N} R_i^2$$

It can be shown that this sum is minimised when the coefficients c_j satisfy a system of linear equations known as the *normal equations*, given by

$$P_{kj} c_j = q_k \tag{5.7}$$

where $\mathbf{P}$ is an N by N matrix and $\mathbf{q}$ is a vector, given by

$$P_{kj} = \sum_{i=1}^{M} B_k(x_i) B_j(x_i) \tag{5.8}$$

$$q_k = \sum_{i=1}^{M} B_k(x_i) y_i \tag{5.9}$$

These equations are in exactly the right form for solution by ELIM (or DECOMP and SOLVE), and in simple cases this will be a perfectly adequate method. In the simplest case of linear regression, the matrix $\mathbf{P}$ is only a 2 by 2 matrix and the solution can be programmed directly; this is done in the next section.

However the method of least squares is not without difficulties, as even when N is fixed and the basis functions chosen, the conditions may be met by more than one set of coefficients c_i. When this happens, or is close to happening, the matrix $\mathbf{P}$ can have a high condition number (see Section 5.1) implying erratic behaviour in the solution. It is important to use a method which can deal with this and give a warning message to the user. A method based on routine SVD (as mentioned in the previous section) is ideal for this purpose. It is described in Section 5.4.4.

⟩ 5.4.2   *Linear regression*

Filename: `4-1REGR.PAS`
To run the working example see *Guide to running the software.*
The listing is also given in this section.
Linear regression is a special case of the general method discussed in the previous section, with basis functions given by

$$B_1(x) = 1, B_2(x) = x \tag{5.10}$$

The matrix P_{kj} is now a 2 by 2 matrix whose elements can be found from equation (5.8):

$$A = P_{11} = \sum_1^M 1 = M \tag{5.11}$$

$$B = P_{12} = P_{21} = \sum_1^M X_i \tag{5.12}$$

$$C = P_{22} = \sum_1^M X_i^2 \tag{5.13}$$

The elements of **q** on the right-hand side of (5.7) become from (5.9):

$$E = q_1 = \sum_1^M Y_i \tag{5.14}$$

$$F = q_2 = \sum_1^M X_i Y_i \tag{5.15}$$

and the normal equations have the form

$$\left. \begin{array}{l} Ac_1 + Bc_2 = E \\ Bc_1 + Cc_2 = F \end{array} \right\} \tag{5.16}$$

The solution is given by

$$\left. \begin{array}{l} c_2 = \frac{(AF-BE)}{(AC-B^2)} \\ c_1 = \frac{(E-Bc_2)}{A} \end{array} \right\} \tag{5.17}$$

In the program, there is no need to store all the data values in arrays as the sums A, B, C, E and F can be calculated progressively as the data are read in. To make the program self-contained the example data have been put in CONST arrays, but the code can be easily modified to use input from keyboard or file.

```
{ 4-1 regression.pas                          }
{ -  Fit and draw regression line - RDH 2/6/89 }


PROGRAM Regline;

  USES
    Cit_core, Cit_prim, Cit_wind, Cit_grap,
      Cit_draw, Cit_ctrl, Cit_disp;

  { Set up arrays and data }
  CONST
    NDATA = 20;
  TYPE
    data_array = ARRAY [1..NDATA,
      1..2] OF Citreal;
  CONST
    XY:data_array = ((1.0, 2.7), (1.5, 4.0), (2.8,
      4.8), (4.0, 4.0), (4.2, 5.2), (6.0, 5.5),
      (5.0, 7.0), (7.0, 5.0), (7.5, 5.0), (8.0,
      5.8), (8.5, 7.0), (9.0, 7.0), (10.0, 6.0),
      (9.5, 5.8), (9.2, 6.5), (1.0, 3.2), (3.0,
      4.5), (5.1, 5.0), (6.8, 5.8), (8.0, 6.9));

  {window variables }
  VAR
    graph_w, results_w : Citwindow;


  VAR
    i, n : integer;
    a, b, c, e, f, k, x, y, c1, c2 : Citreal;
```

```
BEGIN

  { define working windows }
  Define_window (graph_w, 1, 1, 80, 17, CITWHITE,
    CITBLACK, CITBROWN, CITBLACK, PIPYELLOW,
    AXISRED, MARGINS, False);
  Define_window (results_w, 1, 18, 80, 23,
    CITWHITE, CITBLACK, CITBROWN, CITBLACK,
    CITWHITE, CITBLACK, MARGINS, False);

  Graphics_mode;

  Open_window (graph_w, '');
  { open windows }
  Open_window (results_w, 'Results');

  { Plot data }
  Graph_window (graph_w);
  WITH Graphparam DO BEGIN
    xlo := 0.0;
    xhi := 10.0;
    ylo := 0.0;
    yhi := 10.0;
  END;
  Set_scales;
  Draw_axes;
  Label_axes;
  Draw_data1 (@XY, @XY [1, 2], NDATA, 2, PLUS,
    CITYELLOW, False);

  { Initialise sums for P-matrix }
  a := 0.0;
  b := 0.0;
  c := 0.0;
  { Initialise sums for q-vector }
  e := 0.0;
  f := 0.0;
```

```
{ Scan data and form sums }
FOR i := 1 TO NDATA DO BEGIN
  x := XY [i, 1];
  y := XY [i, 2];
  a := a + 1.0;
  b := b + x;
  c := c + Sqr (x);
  e := e + y;
  f := f + x * y
END;

{ now solve Normal Equations }
k := b / a;
c2 := (f - k * e) / (c - k * b);
c1 := (e - b * c2) / a;

{ and draw the line }
Gp_graph_color (CITWHITE);
x := Graphparam.xlo;
Moveabs (x, c1 + c2 * x);
x := Graphparam.xhi;
Drawabs (x, c1 + c2 * x);

Display_real (results_w, 1, 1, 10, 'c1 = ', ' ',
  c1);
Display_real (results_w, 1, 2, 10, 'c2 = ', ' ',
  c2);

Pause;
Quit
END.
```

⟩ *5.4.3   Polynomial fitting using normal equations*

Filename: 4-2POLY.PAS

In this section the least squares method of curve fitting is applied with a polynomial, and the normal equations (equations (5.7)) are solved using routines DECOMP and SOLVE (see Section 5.3).

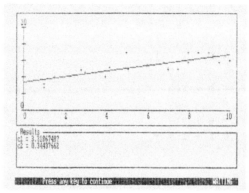

Figure 5.2 A display resulting from program 4-1REGR

The basis functions now are just powers of x. A polynomial with N coefficients has highest degree $(N - 1)$, and the normal equations (see equations (5.7), (5.8) and (5.9)) are given by

$$P_{kj} = \sum_{1}^{M} x^{k+j-2} \qquad (5.18)$$

$$q_k = \sum_{1}^{M} y_i x^{k-1} \qquad (5.19)$$

The matrix $\mathbf{P}$ is symmetric (elements with the subscripts interchanged are equal, $P_{jk} = P_{kj}$) and also any elements for which $(k + j - 2)$ is equal have the same value. As the highest value of each subscript is N, the elements of $\mathbf{P}$ consist of sums of the powers of x_i up to $2N - 2$. The vector $\mathbf{q}$ holds sums of the products of each y_i and powers of x_i.

```
{ 4-2 polynomial.pas                                      }
{ -  Least Squares polynomial fit - RDH 20/3/89 }


PROGRAM Polyfit;
```

```
USES
  Cit_core, Cit_prim, Cit_wind, Cit_grap,
    Cit_matr, Cit_draw, Cit_ctrl, Cit_disp,
    Cit_edit;


{ Set up arrays and data }
CONST
  MDATA = 10;
  MAXNCOEFFS = 8;
TYPE
  data_array = ARRAY [1..MDATA,
    1..2] OF Citreal;
  p_matrix = ARRAY [1..MAXNCOEFFS,
    1..MAXNCOEFFS] OF Citreal;
  coeff_vector = ARRAY [1..MAXNCOEFFS] OF
    Citreal;


CONST
  XY:data_array = ((0.1, 8.8), (0.8, 7.4), (2.2,
    5.2), (3.7, 3.3), (4.5, 2.5), (5.2, 2.0),
    (6.0, 1.4), (7.0, 1.2), (8.1, 0.91), (9.5,
    1.3));


VAR
  a : p_matrix;
  c : coeff_vector;
  aptr, cptr : Arrayptr;


{window variables }
VAR
  graph_w, results_w, coeffs_w : Citwindow;
```

```
VAR
  i, j, n, p : integer;
  sum, cond, det : Citreal;


{$F+}
  { Evaluate a polynomial of degree N-1 }
  { whose coeffs are stored in C[1..N]. }
  FUNCTION poly (xx : Citreal) : Citreal;
    VAR
      k : integer;
      sum : Citreal;
  BEGIN
    sum := 0.0;
    FOR k := n DOWNTO 1 DO
      sum := xx * sum + c [k];
    poly := sum
  END;
{$F-}


BEGIN

  { define working windows }
  Define_window (graph_w, 20, 1, 80, 20, CITWHITE,
    CITBLACK, CITBROWN, CITBLACK, PIPYELLOW,
    AXISRED, MARGINS, False);
  Define_window (coeffs_w, 1, 1, 19, 20, CITWHITE,
    CITBLACK, CITBROWN, CITBLACK, CITWHITE,
    CITBLACK, MARGINS, False);
  Define_window (results_w, 1, 21, 80, 24,
    CITWHITE, CITBLACK, CITBROWN, CITBLACK,
    CITWHITE, CITBLACK, MARGINS, False);


  Graphics_mode;
```

```
  cptr := @c;
  Solve (n, MAXNCOEFFS, aptr, cptr);
  FOR i := 1 TO n DO BEGIN
    Display_integer (coeffs_w, 1, i, 2, 'C', '=',
      i);
    Display_real (coeffs_w, 6, i, 12, ' ', ' ',
      c [i])
  END;

  { and draw the polynomial curve }
  WITH Graphparam DO
    Draw_curve1 (@poly, xlo, xhi, 128, CITWHITE,
      False);

  Pause;
  Quit
END.
```

```
Number of coeffs=2  (max no.=8)
Condition number = 145.61774
     Determinant = 854.89

Coefficients
C1 =   7.343920504
C2 =  -0.837138111

Number of coeffs=3  (max no.=8)
Condition number =  15672.431
     Determinant = 583517.05

Coefficients
C1 =   8.971381065
C2 =  -1.999324814
C3 =   0.125156633

Number of coeffs=4  (max no.=8)
Condition number = 1663370.7
     Determinant =       2.13593E8

Coefficients
C1 =   8.964042088
C2 =  -1.986210632
C3 =   0.121584119
C4 =   0.000247016
```

The data for this example were generated from the formula $y = 9 - 2x + \frac{1}{8}x^2$ with some random variation. The results for three and four coefficients have recovered the original values quite well. But the condition number of the matrix $\mathbf{P}$ is high even with three coefficients, showing that this method (that is, solving the normal equations) is not satisfactory. The condition number has nothing to do with goodness of fit; with two coefficients (linear regression) the condition number is low, but the regression line fits the data poorly compared to the cases with three and four coefficients.

$\rangle$ 5.4.4   General method of least squares

Filename: `4-3SVFI.PAS`

The problems associated with a high condition number can be overcome by a method based on singular value decomposition (see SVD, Section 5.3).

In curve fitting, the aim (see Section 5.4.1) is to find a function $y(x)$ such that $y(x_i)$ equals the observation y_i as closely as possible for each i. Using $\approx$ for 'equals as closely as possible', equation (5.5) gives

$$y_i \approx \sum_{j=1}^{N} c_j B_j(x_i), \ i = 1, 2, ..., M \tag{5.20}$$

This may be written in the form

$$A_{ij} c_j \approx y_i \tag{5.21}$$

where $A_{ij} = B_j(x_i)$. $\mathbf{A}$ is then an M by N matrix called the *design matrix*. If $M = N$ (number of observations = number of coefficients) and the basis functions were properly chosen as all independent, the equations (5.21) above would just be a set of linear equations and solved as in Section 5.1. All the observations could then be matched exactly. However this would not, in general, be possible when N is less than M. This is what interests us now.

The SVD method re-codes the data and the coefficients so that equations (5.21) are transformed, and now, for example with $N = 3$ and M

much larger, they look like

$$
\begin{pmatrix}
s_1 & 0 & 0 \\
0 & s_2 & 0 \\
0 & 0 & s_3 \\
0 & 0 & 0 \\
0 & 0 & 0 \\
\vdots & \vdots & \vdots
\end{pmatrix}
\begin{pmatrix}
c_1' \\
c_2' \\
c_3'
\end{pmatrix}
\approx
\begin{pmatrix}
y_1' \\
y_2' \\
y_3' \\
y_4' \\
y_4' \\
\vdots
\end{pmatrix}
\tag{5.22}
$$

where the s-values (see Section 5.3) are the singular values of the original matrix $\mathbf{A}$.

In this form, the impossibility of getting exact equality with the y_i's becomes obvious, as whatever the values of c_1', c_2' and c_3', the left-hand side will always be zero for all $i > 3$. The best that can be done is to take $c_1 \approx \frac{y_1}{s_1}$, etc., and this is only possible where s_1, s_2 and s_3 are non-zero. The transformed equation also gives a warning if any of the original coefficients are not very well determined. If one of the singular values is very small, the corresponding coefficient could vary considerably without making much difference to the agreement with the right-hand side. This coefficient should be made zero. The threshold for this change depends on the accuracy of the data. If say two figures are regarded as accurate, then it might be sensible to set the threshold at 0.01 of the maximum singular value, and for any singular value less than this the corresponding coefficient could be made zero. However the results of this change should be carefully observed to ensure that the desired effect is achieved.

In this SVD method, it is first necessary to calculate the values of y_i'; only N of these are needed. The formula is given in terms of the matrix $\mathbf{U}$ found by the SVD routine:

$$
y_j' = \sum_{i=1}^{M} y_i U_{ij}
\tag{5.23}
$$

From these we can find the values of the transformed coefficients. These in turn must be inverse-transformed to obtain the values we want. This formula uses the other matrix $\mathbf{V}$ found by SVD:

$$
c_i = \sum_{j=1}^{N} V_{ij} y_j
\tag{5.24}
$$

All these features are contained in the example 4-3SVFI which is not listed here but can be examined on the screen. 4-3SVFI follows the sequence outlined above, reading its data from CONST declarations, evaluating the basis functions and forming the design matrix, obtaining the singular values and then interacting with the user to decide which should be forced to zero. After all this, the coefficients are calculated and then the model is evaluated and compared with the original data. The program works out the maximum difference from the data and the mean square deviation. Finally, a graph of the data points and the fitted curve is drawn. Here is some typical output:

```
Use basis functions 1 to 3    (max 8)
No. of data points = 10

Singular values:
133.3868523     1.192419492     4.802691445

Threshold for singular values: 0.001

Coefficients are:
8.971381066     -1.999324814     0.125156633

Comparison of model with data:
I    X[i]       Y[i]        model
1    0.1        8.8         8.772700151
2    0.8        7.4         7.452021459
3    2.2        5.2         5.178624578
4    3.7        3.3         3.287273561
5    4.5        2.5         2.508841224
6    5.2        2           1.959127393
7    6          1.4         1.481070976
8    7          1.2         1.108782394
9    8.1        0.91        0.988376778
10   9.5        1.3         1.273181482

Root mean square = 0.0525114
Max deviation = 0.912176
```

If the program is run again with $N = 4$, it will be found that the singular values are spread over a far greater range. It turns out that for a sensible fit it is necessary to retain all the coefficients, none being forced to zero despite the fact that the accuracy of the data is no greater. This serves to emphasise the importance of checking the results.

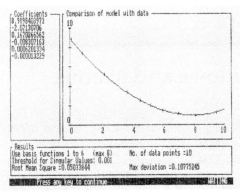

Figure 5.3 A display resulting from program 4-3SVFI

⟩ Chapter 6

⟩ Windows and Menus

For anyone writing a program on a PC, it soon becomes apparent that some degree of organisation is necessary to administer the graphical and numerical tasks that accumulate. There are four units at Level 2 which help do this. They are `Cit_ctrl`, `Cit_menu`, `Cit_disp` and `Cit_edit`. They provide facilities for

- **control** of the flow of your program,
- **menus** for the selection of options,
- **display** of text and numerical values,
- **editing** of text and values.

It is good practice always to write a program with a user in mind. Imagine that the person running your program is on the other side of a glass in a suitably sound-proofed room! The computer screen and keyboard are all there is to provide the interface between you, the program writer, and this other, quite separate person, the program user. The information you display, and the keys the user presses, are likely to be the only means of communication. For this reason some thought has to be given to how information is presented and collected.

To be effective, screens should be designed with an overall plan in mind. Before writing a program, it is worth spending a few minutes thinking through the dialogue that will take place when the user sits at the machine. The user will have to choose options, to specify data, to know what is happening and to see results. It should be clear what information is being presented and asked for on the screen. To a large extent a program will be judged on its appearance and apparent functionality.

The Level 2 routines are meant to help. They can be used without getting too involved in the detail of computing. If you want to know about the detail, this is discussed at length later, in Chapter 8. But for the present, consider the routines and their ability to handle the communications.

212

〉 6.1   Windows

The Level 2 routines operate in a windows environment. A *window* is simply an area of screen which for a time is allocated to communicating about something specific. When a program runs there will be some things to ask the user and some things to tell the user. It is a good idea always to start with a list, preferably organised by function, of the different items that are to be communicated when the program runs. You will then be able to plan the screen layout, allocating windows to these items.

For example, in a program that draws a curve y against x, where

$$y = f(x) = e^{ax} \cos bx$$

there might be a 'heading window', which states what the program will do (e.g. "Plot y=exp(ax) cos(bx)"), a 'parameter window', in which values of a and b, and the x-range $[xlow, xhigh]$ are specified, and a 'graphics window' where the graph is drawn.

These three windows need to be named. Suppose here we adopt the names **heading_w**, **parameter_w** and **graphics_w**. The windows can then be declared in the program code. There is a type **Citwindow**, which is defined in the Toolkit core, and is reserved for this purpose. Thus, the program statement

```
VAR heading_w, parameter_w, graph_w: Citwindow;
```

would be sufficient to declare that windows exist as required.

〉 6.1.1   *Specimen windows*

A user window must be defined before it is used. The computer has to be told where the window is to be placed and what colours are to be used. Normally this will be one of the first operations when a program is executed. There are routines in the Level 1 unit **Cit_wind** which can do this at whatever level of detail is required. However, at Level 2 the specification can be bypassed by copying, or directly using, one of the predefined *specimen* windows. This is convenient, as decisions on positions and colours can be left to the Toolkit (at least initially).

For example, the Pascal statements:

```
heading_w:=Titl_w;
parameter_w:=Edit_w;
graph_w:=Draw_w;
```

could be used to copy details from specimens `Titl_w`, `Edit_w` and
`Draw_w`.

The specimen windows available in the Toolkit are as follows:
`Titl_w` in unit `Cit_ctrl` (for titles)
`Resp_w` in unit `Cit_ctrl` (for short messages)
`Menu_w` in unit `Cit_menu` (for option menus)
`Disp_w` in unit `Cit_disp` (for text displays)
`Edit_w` in unit `Cit_edit` (for data editing)
`Draw_w` in unit `Cit_draw` (for graph drawing)

The function of the individual specimen windows should be clear anyway from the names. They are not used for any other purpose, and are
not altered by the Toolkit. In fact, the names are declared within the
units as global variables of type `Citwindow`. They may therefore be
altered by the user, and may even be adopted directly as user windows.

) 6.1.2   Preset windows

In the above example you might have preferred a more elaborate environment. For example, you could have included instruction for the user
(e.g. "Press any key to continue" or "Enter value and press return"), advice on the program status (e.g. "Waiting" or "Running"), and control
options (e.g. "Repeat" or "Quit").

If so, you might find the *preset* windows useful. Like specimen windows these are already declared as variables of type `Citwindow`, and
have been defined by default. They may be relocated using the Level 1
routines described in Chapter 8. However, preset windows are also used
within the Toolkit, and they should only be used for the designated
purpose.

The full list of preset windows is as follows:
`Status_w, Option_w, Bar_w, Instruct_w,`
`Error_w, Help_w` all of which are in unit `Cit_ctrl`

For our example, the window **Status_w** conveys program status, the window **Instruct_w** carries instructions about which keys to press, and the window **Option_w** contains options which appear in a rolling menu.

Again, the function of these windows is fairly clear from their names.

⟩ 6.1.3   Opening and closing windows

The routines **Open_window** and **Close_window**, which are contained in the Level 1 unit **Cit_wind**, are used to activate windows. The action of 'opening' a window grabs the designated area of screen. Text and graphics written to the window are placed in this area. When the window is 'closed' the screen area is left for another purpose. A typical session, in which data are edited, might have the form:

```
Open_window(Edit_w,'Edit data');
.. edit routines from unit Cit_edit ..
Close_window(Edit_w);
```

The text "Edit data" would be displayed in the margin of the edit window.

⟩ 6.1.4   Writing to a window

In the Toolkit there are a number of ways in which text can be placed in a window. A range of facilities is contained in the Level 1 unit **Cit_wind** and a full description of the subtleties can be found in Chapter 8. An example found there is the routine **Write_string**, the action of which can be understood from the following model:

```
Write_string(Disp_w,3,1,'DETERMINE FLOOR LOADING.');
```

This writes the string **DETERMINE FLOOR LOADING.** in the window **Disp_w** starting at the third column in the first row. Columns and rows are counted from the top-left corner of the window.

These facilities provide a basis on which to work. At Level 2 predefined windows are allocated to the most common categories of information. The routines can be seen in action if you run the demonstration program named **DEM_WIND.PAS**, found in the directory **EXAMPLES\CHAP6**.

This program displays all the specimen and preset windows, and shows how they fit together.

It is apparent from the demonstration program that some windows come and stay on the screen, while others come and go away restoring the screen to its previous state. In fact, window variables can be set for either option. The detail of adjustment is described in Chapter 8. However, as is seen in the demonstration, the predefined settings vary from window to window. There is a reason for this. There are four windows which cover the screen and provide a background, and there seems no reason to save what lies behind them. Everything else has a temporary life. On this basis the default settings cause:

Status_w, Instruct_w, Disp_w, Titl_w to come and stay

Option_w, Bar_w, Error_w, Help_w,
Resp_w, Menu_w, Edit_w, Draw_w to come and go

Level 2 routines will often send messages to the preset windows on their own initiative, for example, when reporting an error or reading a character at the cursor. If this is annoying, it can be suppressed. Windows can be disabled at any time by resetting the window 'type'. For example, the window Status_w can be disabled by including the program statement:

    Status_w.wtype:=NULL;

In summary, the purpose of a window is to carry specific information. Communications with the user should be planned. At Level 2 this can be done using specimen windows and preset windows which have parameters set by default.

⟩ 6.2   Control

Routines in the unit Cit_ctrl provide basic facilities for the user interface.

⟩ 6.2.1   *Status*

Routine name:	Set_status
Kept in unit:	Cit_ctrl
Purpose:	Send message to status window Status_w

The status window provides a quick indication of what the computer is doing. It is opened and a message written. Typical messages are "Waiting", "Reading" and "Running".

Signature:

```
PROCEDURE Set_status(s: Linestring);
```

Input parameters:

Call:	Set_status(s);
s:	Status message

It is good practice to set status and then clear it as soon the reported condition finishes. Sending a null string clears the window.

This routine is frequently called from within other Toolkit routines and the messages then displayed are taken from strings held in the following global variables:

Variable	Default contents
status_waiting	WAITING
status_running	RUNNING
status_computing	COMPUTING
status_reading	READING
status_searching	SEARCHING
status_error	ERROR
status_writing	WRITING

) *6.2.2 Wait*

Routine name:	**Wait_for_keypress**
Kept in unit:	**Cit_ctrl**
Purpose:	Wait a specified time
	for a key to be pressed

Quite often it is required to wait for the user to respond with a single keypress. This routine waits until a specified time has elapsed or until any key is pressed. If a key is pressed, codes are set to indicate which key. The status window is set while waiting.

Signature:

```
PROCEDURE Wait_for_keypress(secs      : Citreal;
                            VAR curcode: integer;
                            VAR keychar: char);
```

Input parameters:

Call: Wait_for_keypress(secs,code,ch);
secs: Wait time in seconds

Output parameters:

code: Integer code for key pressed
ch: Character for key pressed

Values of the variables code and ch are set according to the following table. (A fuller version is in Chapter 8.) The global integer variable Last_key and the Boolean flags Escape and Break are also set. The flags are set **true** only if the respective keys are pressed:

ch	*indicates*	code	Last_key
#$FF	cursor key	numerical value of key	scan code
#$FE	function key	number of function	scan code
#$FD	ALT key	scan code	ASCII char.
#$FC	other special	scan code	scan code
other	character	0	ASCII char.

Wait_for_keypress does not echo the character it receives on the screen.

) *6.2.3   Instructions*

Routine name: Instruct_banner
Kept in unit: Cit_ctrl
Purpose: Send message to instruction window Instruct_w

Instructions in the window Instruct_w tell the user what should be done next. Normally, when a message is sent to the instruction window, another message will be sent to the status window. A typical situation is seen in the following code:

```
Set_status(' WAITING');
Instruct_banner(' Press keys A..D to select option');
Wait_for_keypress(60.0,code,ch)
Instruct_banner('');
Set_status('')
```

This code allows 60 s for a reply.

Signature:

> PROCEDURE Instruct_banner(lline: String);

Input parameters:

Call: Instruct_banner(lline);
lline: Instruction string

The instruction window is opened and a string written. In the string the characters { and } have a special interpretation. They switch highlighting on and off. Control characters can also be included through use of the ^ character. ^A represents $\boxed{\text{Ctrl}}/\boxed{\text{A}}$, ^B represents $\boxed{\text{Ctrl}}/\boxed{\text{B}}$, and so on. The instruction window is cleared if lline is an empty string.

This routine is sometimes used within the Toolkit; the messages then displayed are those contained in strings stored in the global variables instruct_pause, instruct_help, instruct_roll and instruct_bar. These variables are set by default to appropriate messages and may be altered by the user.

⟩ 6.2.4   *Pause*

Routine name: **Pause**
Kept in unit: Cit_ctrl
Purpose: Wait indefinitely for a user response

An indefinite pause gives the user an opportunity to take time reading the screen and thinking.

Signature:

> PROCEDURE Pause;

While waiting, the instruction and status windows are set. On return the global variables Last_key, Escape and Break are modified as indicated above.

⟩ 6.2.5   *Errors*

Routine name: **Report_error**
Kept in unit: Cit_ctrl
Purpose: Send message to error window **Error_w**

As described in Chapter 3, there are global variables dedicated to handling error messages. The Boolean variable `Errorflag` is set **true** when an error occurs, and indicates that an error code has been set in the integer variable `Errorcode`, with an explanation in the string variable `Errorstring`. The routine `Report_error` provides a mechanism for reporting errors. Typically, the code would appear as follows:

```
IF Errorflag THEN Report_error(Errorstring);
```

Signature:

```
PROCEDURE Report_error(s: Linestring);
```

Input parameters:

Call:	`Report_error(s);`
s:	Error message

The error string is centralised within the window. By default, the message stays there for 2.5 s. This value is taken from the global variable `Report_wait_time` (of type `Citreal`), which can be changed by the user. If it is set to zero, then the wait is indefinite. Pressing the ⃞Esc or ⃞Break keys cancels an error report immediately.

The routine also leaves a message in the status window. The message is contained in the global variable `status_error`. By default the message is ERROR. Thus even if the error window itself is disabled, either because the user has disabled it, or because there is simply not enough room on the heap to save the background to the error window, there is still some indication of an error.

The error window has a privileged status. Whereas in general there is a certain amount of memory reserved for tasks other than saving backgrounds to window images (the actual amount being specified in the global variable `non_window_reserve`), this memory *is* available for an error window. Therefore it is quite likely that the error window can be opened for all situations except when numerical calculations require array space to the limits of memory.

⟩ **6.3   Help**

There is a *help* facility within the unit `Cit_ctrl`. This can be used on its own or in conjunction with the option and menu routines.

The help facility is based on a set of ASCII text files prepared on a word-processor or in the Turbo Pascal environment. The files contain help messages.

The help messages can be read as a block into a message list maintained in memory. The routine **Read_help** will do this, and should be called at the beginning of the program. Alternatively, the action of reading help files can be postponed until the user calls for help. In this case the individual messages are extracted one-at-a-time and are added to the message list.

Once placed in the list a message can be read and placed on the screen with speed, the messages being identified by a string marker which is used as a reference.

The help messages can be considered grouped according to *categories* of advice appropriate for different circumstances. The category will usually be indicated by the first few characters in the reference marker. In a program, a category of help can be 'opened' by a call to the routine **Open_help**. At the same time the name of the help file which provides the source, and the characters that distinguish the reference marker are specified.

Once a category of help has been opened, help messages can be called onto the screen using the routine **Call_help**. This routine performs a search first through the list in memory, and then through the file named for the current category, until the message is found. It displays the message in the help window set for the current category. The window is opened if it is not currently open. It can be closed later by a call to the routine **Wipe_help**. Thereafter, calls to the routines **Call_help** and **Wipe_help** will display further messages.

Several help categories can be open at the same time, the most recent being the one currently active. Help categories can be closed by a call to the routine **Close_help**. Then the next most recently opened category is resumed.

There is a global Boolean variable **Help_on** which must be set **true** for the help facility to operate. It is set **true** automatically when a category is opened. Some of the Level 2 routines call help directly and, if this becomes a nuisance, then **Help_on** should be set **false**.

⟩ *6.3.1   Format of help files*

A help file comprises lines of ASCII text which form help messages, following one after the other. The search *markers*, by which the individual messages are recognised, are character strings, each preceded by a character $ and placed on a line by themselves. The message then starts on the next line and continues until a # character is found.

Typically, the marker strings end in a character taken from the sequence of capital letters A,B,C,D,.... Extra characters are added as the messages are taken from further up the hierarchy.

For example, suppose help is arranged at two levels, the first being available when choosing options from a menu, and the second giving advice as the user steps through a process of computation. The help file would be arranged as follows:

```
$MenuA
This option  provides ...
#
$MenuAA
Firstly, calculate ...
#
$MenuAB
Now calculate ...
#
$MenuB
This option provides ...
#
$MenuBA
Firstly, calculate ...
#
$MenuBB
Now calculate ....
#
```

There are two parts to each marker, a *root* and an *extension*. In this example, **MenuA** and **MenuB** are markers at the top level, and have a root **Menu** with extensions **A** or **B**. At the next level the root will be either **MenuA** or **MenuB**, according to which option is chosen, and the extensions would be either the letter A or the letter B.

There can be many help files, and many categories of message contained in each file. However, help for one category can only be stored in one file.

) 6.3.2   *Maintaining a list of messages*

A message list is maintained in the memory. If access to help messages is to be quick, then it may be better to read all, or a significant number of, messages from the file when the program begins to run.

Routine name: `Read_help`
Kept in unit: `Cit_ctrl`
Purpose: Read a file of help messages

The file should comprise text arranged as described above.

Signature:

> `PROCEDURE Read_help(filename: Linestring);`

Input parameters:

Call: `Read_help(fname);`
fname: Name of help file (default extension `.HLP`)

Several files can be read into the message list. If it happens that a message is not present when required then the file is searched and the individual message is extracted.

The information stored in the list can be deleted, and memory freed, by a call to the following routine.

Routine name: `Free_help`
Kept in unit: `Cit_ctrl`
Purpose: Clear all help messages from memory

Signature:

> `PROCEDURE Free_help;`

$\rangle$ *6.3.3   Opening and closing categories of help*

Routine name: `Open_help`
Kept in unit: `Cit_ctrl`
Purpose: Open a category of help

A help stack is maintained in the memory listing all help categories currently open. When a new category is opened a note is made of the filename and the root of the search marker. The help stack is then pushed up one level and a new window chosen. The position and colours of the new window are taken from the variable `Help_w` as it is set at that time. This new window will be used to display help for the new category.

Signature:

```
PROCEDURE Open_help(filename,title,root: Linestring);
```

Input parameters:

Call:	Open_help(fname,title,root);
fname:	Name of the help file
title:	Title displayed in margin of help window
root:	Default root on which to base searches for markers

The default extension of the help file name is .HLP. At the second and higher levels, the string root is added to the existing root, i.e. the one that was present at the last level. Thus, the root accumulates characters as help moves through the branches of the overall help structure.

Routine name:	Close_help
Kept in unit:	Cit_ctrl
Purpose:	Close a category of help

After a category of help has been explored it will be closed. The corresponding help window will also be closed, and the search markers will resume the setting that applied before that category was opened.

Signature:

```
PROCEDURE Close_help;
```

The list of messages is not removed from memory, and the messages can be used on another occasion if their root is named in other help categories.

⟩ *6.3.4   Calling and fetching help messages*

Routine name:	Call_help
Kept in unit:	Cit_ctrl
Purpose:	Calls a help message in the current category

This routine searches for the help message corresponding to the specified marker. If it is not already open, the help window for the current category is opened, and the message displayed.

Signature:

```
PROCEDURE Call_help(root,extn: Linestring);
```

Input parameters:

Call:	`Call_help(root,extn);`
root:	Root used when searching for a marker
extn:	`Search marker:=root+extn`

When displaying text the { and } characters have special significance, acting as switches which turn highlighting on and off. There are two ways by which the message can be referenced. If the parameter **root** is set to the null string, then the category's root is adopted. Thus the messages within a category (for example, those corresponding to options in a menu) are distinguished by the string in the parameter **extn**. Alternatively, if the parameter **root** contains a string, then this is used with the parameter **extn**.

Routine name:	**Fetch_help**
Kept in unit:	**Cit_ctrl**
Purpose:	Retrieve the text for a help message

The function **Fetch_help** operates in the same way as the procedure **Call_help** except that it does not go as far as writing to the help window. Instead it returns a pointer to the help message stored in memory. This pointer can be used by Level 1 routines which will read the text directly.

Signature:

```
FUNCTION Fetch_help(root,extn: Linestring) : Textptr;
```

Output parameters:

Fetch_help:	Returns a pointer, of type **Textptr**

⟩ *6.3.5   Wiping help messages*

Routine name:	**Wipe_help**
Kept in unit:	**Cit_ctrl**
Purpose:	Close the current help window, but keep the help level open

When help is 'wiped' the current window is closed. It will be reinstated automatically when help is next called.

Signature:

```
PROCEDURE Wipe_help;
```

The demonstration program **DEM_HELP.PAS** which can be found in the directory **EXAMPLES\CHAP6** reads a help file **DEM_HELP.HLP** which is constructed using the above structure.

) 6.4   Options

You often want the user to select from a list of options. An option menu will achieve this. Two simple forms are available in the unit **Cit_ctrl**; they provide a rolling menu and a one-line bar menu.

) 6.4.1   *Rolling options*

If option strings are long and screen space is limited, then it may be better to roll the options through a small one-line window.

Routine name:	**Roll_option**
Kept in unit:	**Cit_ctrl**
Purpose:	Presents options in the form of a rolling menu

Help can be made available; it depends on whether a help category is open. If it is, then, when F1 is pressed, a help message is 'called' to match a marker string made up from the current 'root' with an 'extension' of a capital letter A,B,C,..., chosen according to the option number in the menu. The first option gives extension **A**, the second option gives extension **B**, and so on.

Signature:

```
PROCEDURE Roll_option(prestring : Linestring;
                      optionlist: STRING;
                      col,row   : integer;
                      VAR opt   : integer);
```

Input parameters:

Call:	Roll_option(s,list,col,row,opt);
s:	String that precedes the option box
list:	List of options
col, row:	Position of string s within the window Option_w
opt:	Default option number

The option strings are listed in the parameter list separated by a $ character. For example, the routine call might be

```
Roll_option('Choose:','Opt 1$Opt 2$Opt 3',1,1,opt);
```

The option window is highlighted and set to the maximum width necessary for any option in the list.

Output parameters:

opt:	Chosen option

If Esc or Break is pressed, the corresponding flag is set and a zero value is returned in the parameter opt.

⟩ *6.4.2   Bar option*

If the option strings are short, then it may be possible to list them side-by-side across the screen in a one-line bar menu.

Signature:

```
PROCEDURE Bar_option(prestring : Linestring;
                     optionlist: STRING;
                     col,row    : integer;
                     VAR opt    : integer);
```

Input parameters:

Call:	Bar_option(s,list,col,row,opt);
s:	String that precedes the option bar
list:	List of options
col, row:	Position of string s within window Bar_w
opt:	Default option number

The parameters are set in a similar way to the parameters for the routine `Roll_option`. A separator, the string held currently in the global variable `bar_division`, is included anyway between the options, but extra spaces can still be added to the parameter string `list`. The option string that appears in the string `list` will be highlighted when selected.

Output parameters:

`opt`:   Chosen option

If a category of help is open, then help messages will be called after F1 is pressed. Options can be selected by pressing Enter and also by pressing a key for the first non-space character of the option string. If there are two or more items with the same letter, then the next option to the right is chosen. If Esc or Break is pressed, the corresponding flag is set and a zero value is returned in the parameter `opt`.

⟩ **6.5   Menus**

Sometimes program options become an important feature. For example, a program might be designed to give an appreciation of methods that run parallel to each other. In this case the options would be displayed as a menu, and selections made at the user's discretion. The program support then has to be more elaborate. Support for this is provided by the unit `Cit_menu`.

Though the menu facilities are comprehensive, they operate in a very simple way. Any window can be a menu; any non-empty line in which the first character is a space can be an option. There are facilities to move a menu bar up and down through the options. If the first character on a line is not a space then that line is skipped, and the line can form a sub-heading or division in the list of options.

By convention, menu options are selected by pressing the Enter key. The option value returned is the line number in the window. It is left to the program writer to interpret this number as the value corresponding to a particular selection.

If the keys Esc or Break are pressed, a zero value is returned and the respective global variable is set **true**. Also, if, while moving up and down, the left or right cursor arrows are pressed, a zero option is again reported. This time, though, the global variable **slideout** is set **true**. If you do not want to 'slide out' of the menu, then simply re-enter it.

Otherwise, it is possible to drop back to other menus in a larger menu structure.

A specimen window **Menu_w** is provided. There is space for 10 options, but this can be changed using the Level 1 routine **Locate_window**.

Help can be accessed within the menu and will relate to the line being highlighted in the option list. This facility will operate if a help category has been opened. Then, when F1 is pressed, a message is sought based on a marker made up from the current root with an extension taken to be the **opt**th capital letter A, B, C, ... If help is not required, but a help category has been opened for some other reason, then help can be disabled by setting the global variable **help_on** to be **false**.

In addition, there is a global Boolean variable **auto_help_on** which, when set **true**, indicates that a help message is to be called while the menu is being displayed. This message can contain general guidance on the options available in the overall menu. It will be the message in the help file identified by the marker which matches the current root.

When selections are being made from the menu, appropriate instruction and status messages are displayed. The instruction issued is the string contained in the global variable **instruct_menu**, possibly enhanced by the message **instruct_help** if help is switched on. The status message is that contained in the global variable **status_waiting**.

) *6.5.1 Open and close menus*

Routine name:	**Open_menu**
Kept in unit:	**Cit_menu**
Purpose:	Open a menu window and write a list of options

The process of opening a menu comprises opening the named window, and copying the options. Also, any general 'auto-help' message is displayed.

Signature:

```
PROCEDURE Open_menu(VAR w    : Citwindow;
                    caption  : Linestring;
                    menulist : STRING);
```

Input parameters:

Call:	`Open_menu(w,capt,list);`
`w`:	Current help window
`capt`:	Caption string displayed in window margin
`list`:	Option string

The option string contains the menu options separated by the character `$`. The leading space character needed for an option is added automatically. However if the option string begins with a character `-`, then it is assumed that the line is to be a division and the space is not added.

It might not always be possible to accommodate all the option strings within the parameter string `menulist`. An alternative approach is to open the window with an empty list and then subsequently write the option text using, for example, the routine `write_string`.

Several menus can be open at the same time, each referred to by the name of its window.

After selecting an option the menu can be left open for future reference whilst, for example, further menus are explored, or it may be closed directly. In the latter case the screen window is closed and any help message that might have remained through use of the `auto_help_on` facility is wiped.

Routine name:	`Close_menu`
Kept in unit:	`Cit_menu`
Purpose:	Close a menu window and wipe any message arising through 'auto-help'

Signature:

```
PROCEDURE Close_menu(VAR w : Citwindow);
```

Input parameters:

Call:	`Close_menu(w);`
`w`:	Current help window

⟩ *6.5.2   Select from menu*

An option can be selected by pressing the [Enter] key when the required line is highlighted, the menu bar being moved using the up and down cursor keys. An option can also be selected by pressing a key that matches the first non-space character on the required line. Thus, it is possible to give the options numbers 1, 2, 3, 4, ..., or letters A, B, C, D, ..., and then make the choice entirely by pressing one key. If two or more lines have the same first character, the next line is chosen beyond the current position of the menu bar.

If the help key $\boxed{\text{F1}}$ is pressed and help is active, then a specific help message, relating to the option on which the menu bar is positioned, will appear. It will disappear as soon as any key is pressed.

Routine name: `Select_from_menu`
Kept in unit: `Cit_menu`
Purpose: Select an option from an open menu window

Signature:

```
PROCEDURE Select_from_menu(w        : Citwindow;
                           VAR option : integer);
```

Input parameters:

Call: `Select_from_menu(w,opt);`
`w`:   Current menu window
`opt`:   Default option

If there is no currently valid option with the default line number, then a search is made for the next acceptable line.

Output parameters:

`opt`:   Number of the row containing the chosen option

) 6.5.3   Rest and wake menu

If, when an option is selected, you become engrossed in an activity to which the option has led, you may wish to push the menu into the background, whilst reminding the user of the menu from which he came. This is called *resting*. The menu is painted dull colours while resting. You can *wake* it, restoring the original colours later.

There are two global variables `dull_color1` and `dull_color2` which can be set to dull foreground and dull background colours, respectively. By default these are the sober light-grey and black.

In the act of resting a window, help messages are wiped from the screen. If the 'auto-help' facility is active, then the menu help window is displayed as soon as the menu is wakened.

Routine name:	`Rest_menu`
Kept in unit:	`Cit_menu`
Purpose:	Paint menu window dull colours

Signature:

```
PROCEDURE Rest_menu(w: Citwindow);
```

Input parameters:

Call:	`Rest_window(w);`
`w`:	Current menu window

If you wake a menu, the original foreground and background colours are restored, and selections can be made once more.

Routine name:	`Wake_menu`
Kept in unit:	`Cit_menu`
Purpose:	Revive the colours in a menu window

Signature:

```
PROCEDURE Wake_menu(w: Citwindow);
```

Input parameters:

Call:	`Wake_menu(w);`
`w`:	Current menu window

⟩ *6.5.4   Restrict menu*

It is often convenient to restrict menu choices from time to time, when particular options are no longer applicable. A menu option can be disabled by placing any non-space character in the first position in a row. By convention a ▌ symbol is placed there and the line is painted the dull colours described above.

   Restrictions can be imposed in one call to the routine **Restrict_menu**.

Routine name:	`Restrict_menu`
Kept in unit:	`Cit_menu`
Purpose:	Restricts menu choice and dulls options

Signature:

```
PROCEDURE Restrict_menu(w     : Citwindow;
                        mask : longint);
```

Input parameters:

Call: Restrict_menu(w,mask);
w: Current menu window
mask: Bit-mask

mask is a parameter of type **longint** in which bits are set to the value
1 if an option is to be excluded. The least significant bit of mask (with
value 1) controls the top line of the menu.

⟩ *6.5.5   Select from sub-menu*

There is a routine which incorporates opening, selecting from and closing
a menu in one call. Typically, it will be used for a sub-menu selection.
The sub-menu appears and disappears as soon as an option is selected.

Routine name: Select_from_submenu
Kept in unit: Cit_menu
Purpose: Display and select from sub-menu

Signature:

```
PROCEDURE Select_from_submenu(w         : Citwindow;
                             VAR option : integer);
```

Input parameters:

Call: Select_from_submenu(w,opt);
w: Sub-menu window
opt: Default option

Output parameters:

opt: Chosen option

The same conventions apply as for the routine Select_from_menu.

⟩ **6.6   Display**

A principal task of communication is the delivery of information to the
user. Normally, a number of windows will be reserved specifically for

the display of values and text. At Level 1, routines like **write_string** offer the service of writing to the screen. Their function is a basic one of exploiting the window conventions; they work with character strings. However, at Level 2, the routines are more concerned with displaying basic mathematical objects, like values and arrays, and text objects.

At the higher level there have to be some assumptions about formats. Level 2 provides a set of facilities which should be acceptable for many situations. They are built on the more detailed Level 1 formatting and window routines. The same conventions apply, but the Level 2 routines have been conceived with the aim of displaying objects in mind. They should therefore prove more convenient.

The display routines are all contained in the unit **Cit_disp**. There are companion facilities for editing objects and these are contained in the unit **Cit_edit** which is described in the next section. The same defaults have been used there. The distinction between the two units is that, whereas the routines in **Cit_edit** allow the user to interact and change values and text, the routines in **Cit_disp** do not. In this section, then, communication is one-way, being strictly screen to user.

It will be noticed that all the display routines send messages to a named window. There is a specimen window **Disp_w** provided, but the programmer is free to nominate any other defined windows for this purpose. The messages begin at a position specified within that window. Furthermore, the display field can be preceded and followed by text strings. These are normally used for explanation. If there is a non-empty prestring, then it is this prestring that starts at the specified position, the display field being placed to the right of the prestring.

There is a global variable **Display_String** defined in the unit. It is of type **Linestring** and contains a copy of the last item displayed as a string. This is useful if the program user is to be reminded at a later stage of the exact form of the current display.

⟩ *6.6.1   Display values*

Single values will be of type **Citreal** or **integer**. In the former case there are three possible formats, *real*, *fixed-point*, and *floating-point*. In all cases there is a field width to specify. For fixed-point formats, the decimal point will be at a specified position in the field. For floating-point formats, it can be at any point in a field full of characters. For real formats, an attempt is made to find a suitable arrangement of characters.

If a real value does not fit the field specification, then the field is left blank. If an integer value does not fit, then the field is expanded.

The Level 1 formatting routines are used to display real values. The **sign** parameter for these options is taken from the default settings of the global variables **format_sign_real**, **format_sign_fixpt** and **format_sign_float** found in unit **Cit_form**. For more detail on formatting options, see Chapter 9.

There are four routines of relevance here. The parameters they require are similar.

Routine name: **Display_real**
Kept in unit: **Cit_disp**
Purpose: Display a value in *real* format

Signature:

```
PROCEDURE Display_real(VAR w         : Citwindow;
                       col,row,field : integer;
                       prestring     : Linestring;
                       poststring    : Linestring;
                       r             : Citreal);
```

Input parameters:

Call: **Display_real(w,col,row,field,pre,post,r);**
w: Current display window
col, row: Position of start of string **pre**
field: Width of display field
pre, post: Strings which precede and follow the display field
r: Value displayed

Routine name: **Display_fixpt**
Kept in unit: **Cit_disp**
Purpose: Display a value in *fixed-point* format

Signature:

```
PROCEDURE Display_fixpt(VAR w      : Citwindow;
                        col,row,
                        field,ndec : integer;
                        prestring,
                        poststring : Linestring;
                        r          : Citreal);
```

Input parameters:

Call: `Display_fixpt(w,col,row,field,n,pre,post,r);`
n: Number of decimal places. (Can be zero)
                Other parameters as for `Display_real`

Routine name: `Display_float`
Kept in unit: `Cit_disp`
Purpose: Display a value in *floating-point* format

Signature:

```
PROCEDURE Display_float(VAR w          : Citwindow;
                        col,row,field : integer;
                        prestring     : Linestring;
                        poststring    : Linestring;
                        r             : Citreal);
```

Input parameters:

Call: `Display_float(w,col,row,field,pre,post,r);`
                Parameters as for `Display_real`.

Routine name: `Display_integer`
Kept in unit: `Cit_disp`
Purpose: Display a value in *integer* format

Signature:

```
PROCEDURE Display_integer(VAR w          : Citwindow;
                          col,row,
                          field        : integer;
                          prestring    : Linestring;
                          poststring   : Linestring;
                          i            : integer);
```

Input parameters:

Call: `Display_integer(w,col,row,field,pre,post,i);`
i: Value displayed
                Other parameters as for `Display_real`

If the integer value is greater than the largest value that might be accommodated by the field width, then the field is automatically enlarged. Thus, setting the field to zero will create a field just wide enough. However, it will also mean that you cannot predict where the line will end.

The string **poststring** always immediately follows the value on the screen.

⟩ 6.6.2   *Display arrays*

An array of real values can be displayed using one routine call. The values are described in Toolkit array objects pointed to by variables of type **Arrayptr**. The assumption is made that the array has at least **ni** rows of columns with **n0** real elements, the first **nj** of which are to be displayed. This provides some scope for displaying row vectors, column vectors and sub-matrices.

The array will be placed on the screen within a sub-area of the named window. The first element is placed at the top-left corner of a specified area. The top-left corner of this area is positioned within the specified window at the col^{th} column along the row^{th} line counting from the top left of the window. The extent of the area is specified as a number of columns, **ncols**, and a number of rows, **nrows**.

Array elements are separated horizontally by one space. If an array is too large to fit in the sub-window, then the text is truncated.

It is quite possible to display any part of the array by changing the array pointer. The pointer value is simply a memory address, and the routine is not told of the significance of that address in relation to the array. So, if the pointer were to be set at the address of any element in the middle of the array, then the display would start from there. Of course, it would be left to the programmer to ensure that the size of the array, as given in values of the parameters **ni** and **nj**, was adjusted accordingly.

If the Boolean parameter **border** is set true, then a count is displayed of the rows and columns in a border within the sub-window. This adds an extra row and five extra columns to the display, leaving less space available for values inside the sub-window.

There are three routines, giving the option of real, fixed-point or floating-point formats in the display.

Routine name:	**Display_array_real**
Kept in unit:	**Cit_disp**
Purpose:	Display vectors and matrices in **real** format within a specified sub-window

Signature:

```
PROCEDURE Display_array_real(VAR w       : Citwindow;
                            col,row      : integer;
                            ncols,nrows  : integer;
                            field        : integer;
                            border       : boolean;
                            a            : Arrayptr;
                            ni,nj,n0     : integer);
```

Input parameters:

Call: Display_array_real(w,col,row,nc,nr,fld,
                                    bord,@A,ni,nj,n0);

w:	Current display window
col, row:	Position top-left corner of sub-window
nc, nr:	Number of columns and rows displayed
fld:	Width of field occupied by each element
bord:	Boolean flag, set **true** if rows and columns are to be enumerated
@A:	Pointer to array
ni, nj:	Number of rows and columns of array displayed
n0:	Row dimension of array A

Routine name:	`Display_array_fixpt`
Kept in unit:	`Cit_disp`
Purpose:	Display vectors and matrices in fixed-point format within a specified sub-window

Signature:

```
PROCEDURE Display_array_fixpt(VAR w     : Citwindow;
                             col,row    : integer;
                             ncols,
                             nrows      : integer;
                             field,
                             ndec       : integer;
                             border     : boolean;
                             a          : Arrayptr;
                             ni,nj,n0   : integer);
```

Input parameters:

Call:	`Display_array_fixpt(w,col,row,nc,nr,fld,`
	`                    ndec,bord,@A,ni,nj,n0);`
ndec:	Number of decimal places displayed
	Other parameters as for `Display_array_real`

Routine name:	`Display_array_float`
Kept in unit:	`Cit_disp`
Purpose:	Display vectors and matrices in floating-point format within a specified sub-window

Signature:

```
PROCEDURE Display_array_float(VAR w     : Citwindow;
                             col,row   : integer;
                             ncols,
                             nrows     : integer;
                             field     : integer;
                             border    : boolean;
                             a         : Arrayptr;
                             ni,nj,n0  : integer);
```

Input parameters:

Call:	`Display_array_float(w,col,row,nc,nr,fld,`
	`                    bord,@A,ni,nj,n0);`
	Parameters as for `Display_array_real`

⟩ *6.6.3   Display range*

It is quite common, for example when drawing the graph of a function, to want to display the range of values being considered. Upper and lower bounds might be held in two variables, $xl = 0.0$ and $xh = 10.0$, say. Suppose it is required to display on screen a statement which says "Values of x in the range $[0.0, 10.0]$".

The routine `Display_range` will display statements of this form. Further, the statement will appear exactly the same as it might have been produced by the `Edit_range` routine described later; the two routines complement each other.

Routine name:	`Display_range`
Kept in unit:	`Cit_disp`
Purpose:	Display a range of real values

Signature:

```
PROCEDURE Display_range(VAR w        : Citwindow;
                        col,row,field: integer;
                        prestring    : Linestring;
                        midstring    : Linestring;
                        poststring   : Linestring;
                        l,h          : Citreal);
```

Input parameters:

Call: Display_range(w,col,row,field,
                                          pre,mid,post,l,h);

w: Current display window
col, row: Position of start of display line
field   Width of value fields
pre, mid, post: Separating strings
l, h: Low and high values in range

In the example quoted, the Pascal statement would be

```
display_range(w,1,1,4,
    'Values of x in the range [' , ',' , ']' ,
    xl,xh);
```

⟩ 6.6.4   Display text

One of the features of the Toolkit is its ability to accept and manipulate multi-line text. Multi-line text is stored as heaped text and referenced by a variable of type **Textptr**.

The situation will commonly arise where arrays and expressions are input in text form, perhaps from the keyboard or from an ASCII file, or where a display is constructed in memory and then examined in sections, as might be the case when constructing a spreadsheet. In these situations there will be text objects which need to be edited and displayed.

The routine **Edit_text**, which is described in detail later in this chapter, can handle input of text from the keyboard. Here we consider how to display the text input from this and possible other sources.

Inasmuch as the text can occupy several lines, a sub-window is specified within the display window. The sub-window is positioned with its top-left corner at the position specified by the parameters **row** and **col**.

Routine name: **Display_text**
Kept in unit: **Cit_disp**
Purpose: Display text objects

Signature:

```
PROCEDURE Display_text(w          : Citwindow;
                       col,row    : integer;
                       ncols,nrows : integer;
                       t          : Textptr;
                       offsetx,offsety : integer);
```

Input parameters:

Call:	Display_text(w,col,row,nc,nr,t,x0,y0);
w:	Current display window
col, row:	Position of start of string **prestring**
nc, nr:	Number of columns and rows in the sub-window
t:	Pointer to the text object
x0, y0:	Position within the text of the top-left corner of the sub-window

When the routine writes to the sub-window it completely refreshes the screen. The effect, if the position parameters x0 and y0 are changed a unit at a time, is to scroll through a virtual screen. Refreshing the screen takes a time depending on the current screen mode. Generally, writing to the screen in text mode is fast, and even quite large sub-windows will scroll smoothly. For graphics modes, though, it might be preferable to attempt scrolling only small windows.

) *6.6.5   Display expression*

One of the mathematical objects which is constructed from text and which is commonly used in the Toolkit is the multi-line expression. It will be stored as heaped text and referenced by a variable of type **Textptr**.

A mathematical expression can be entered using the routine **Edit_expression**, which is described in detail later in this chapter, displayed by the routine **Display_expression**, and compiled and evaluated using routines in the unit **Cit_eval**. Here, we consider displaying the expression.

A sub-window has to be specified within the display window, which is where the expression will be displayed. (It might be only one line deep.)

The sub-window is positioned with its top-left corner at the end of an introductory character string named **prestring**. This string itself starts at a position specified by the parameters **row** and **col**.

Routine name:	Display_expression
Kept in unit:	Cit_disp
Purpose:	Display an expression, complementing the routine Edit_expression

Signature:

```
PROCEDURE Display_expression(w          : Citwindow;
                            col,row     : integer;
                            ncols,nrows : integer;
                            prestring   : Linestring;
                            expr        : Textptr);
```

Input parameters:

Call:	Display_expression(w,col,row,nc,nr, prestring,expr);
w:	Current display window
col, row:	Position of start of string prestring
nc, nr:	Number of columns and rows in the sub-window
prestring:	Preceding string
expr:	Pointer to the expression text

) *6.6.6   Display name*

Displaying names is a relatively simple task and might well be carried out directly using the Level 1 routine write_string. However it is often useful, after editing a name, to repeat a display, keeping the same format. For this reason the routine display_name is included. It complements the routine edit_name which is described later in this chapter.

Routine name:	Display_name
Kept in unit:	Cit_disp
Purpose:	Display a name string, complementing the routine Edit_name

Signature:

```
PROCEDURE Display_name(VAR w         : Citwindow;
                       col,row,field : integer;
                       prestring     : Linestring;
                       poststring    : Linestring;
                       str           : Linestring);
```

Input parameters:

Call:	`Display_name(w,col,row,fld,pre,post,s);`
w:	Current display window
col, row:	Position of start of introductory string
fld:	Width of display field
pre, post:	Strings that precede and follow the name
s:	Name displayed

Like the routine `Edit_name`, this routine can be used quite generally to display small text strings.

) 6.7   Edit

Another important communication task is the collection of information from the user either by entering a new object from the keyboard, or by editing an existing one. We shall call both activities *editing*.

At Level 2 there are routines which provide the service of editing the standard Toolkit objects, namely text, strings, real values, arrays, integers, expressions and names. The routines form a set, contained in the unit `Cit_edit`. All the routines use the same key conventions. Further, the display formats are the same as those in the last section. In fact, for most of them there are complementary routines in the unit `Cit_disp`.

The objects are edited in an edit field defined inside an edit window. There is a specimen window `Edit_w` available in the unit, though, of course, the user is free to define and use any window variable.

The edit field forms a sub-window through which text is viewed. Although the size of the edit field has to be specified in advance, it is not always known exactly how many characters the user will want to type. To resolve this difficulty, the edited text is allowed to scroll sideways in the sub-window. The keys F5, F6, F7, and F8 control the cursor and make it jump through the text.

Whilst editing is active the edit sub-window is highlighted. Editing can be in 'insert' or 'overwrite' mode. The mode is toggled using Ins.

By convention, a default value is offered to the user, even when a new object is entered. This is particularly important when the program is to be part of a demonstration, when the minimum of keystrokes are required. The F9 key can be used to clear the edited text. Pressing F10 restores the default.

Although not allowed by the defaults, it is possible to invoke a *copy* facility similar to that available on the BBC micro. There is a global Boolean variable `allowcopy` which, if set `true`, will allow the cursor to be taken outside the edit field and be placed over other characters on the screen. If Ins is then pressed, characters will be copied from the block cursor to the position of the flashing cursor in the edit field. By default, the variable `allowcopy` is set `false`. When the copy facility is available, the keys F3 and F4 can be used to switch the copy block on and off.

The instruction and status windows are set by the edit routines. If they are not required the respective window variables should be set to the value `NULL` as described at the beginning of the chapter. The Toolkit displays messages held in the global variables `instruct_edit`, `instruct_edit_copy`, `instruct_edit_scroll`, `instruct_array`, `instruct_view` and `instruct_view_array`. The choice of message will depend on whether the copy facility has been activated and whether the edit is applied to a single- or multi-line window. In the latter case the text will 'scroll' up and down.

Help can be switched on whilst editing. A help category should be open, and the variable `help_on` set `true`. As indicated in the instruction window, pressing the key F1 will call help. Subsequently, pressing any key wipes the message.

⟩ *6.7.1   Editing strings and text*

At its most basic, editing is a matter of amending strings of characters on the screen. The routine `Edit_string` provides facilities for entering and editing one line of text.

By way of contrast, the routine `Edit_text` edits multi-line text, and a companion routine `View_text` permits the user to scan through multi-line text without actually changing the characters. The multi-line text can be assembled and manipulated using the Level 1 routines in the unit `Cit_text`, and is stored in memory at a position denoted by the contents of a pointer variable of type `Textptr`. The routine `Edit_text` allows the text to be assembled and changed at the keyboard.

Other Level 2 routines read the multi-line text as if it described arrays or expressions.

Routine name:	`Edit_string`
Kept in unit:	`Cit_edit`
Purpose:	Display and edit a string of characters

Signature:

```
PROCEDURE Edit_string(w          : Citwindow;
                      col,row,len : integer;
                      VAR s       : string);
```

Input parameters:

Call:	Edit_string(w,col,row,len,s);
w:	Current edit window
col, row:	Position of edit field within edit window
len:	Length of edit field
s:	Default string

Output parameters:

s:	Edited string

Pressing ⎡Enter⎤, ⎡Tab →⎤, or ⎡Tab ←⎤ denotes a successful end to the editing. The value of the Level 1 variable **last_key** can be used to determine which key terminated the string. If ⎡Esc⎤ or ⎡Break⎤ is pressed, the parameter **s** is set to the default string, the global variables **Escape** and **Break** being set accordingly.

Routine name:	Edit_text
Kept in unit:	Cit_edit
Purpose:	Edit text objects

Signature:

```
PROCEDURE Edit_text(w                  : Citwindow;
                    col,row,ncols,nrows : integer;
                    var t               : Textptr;
                    offsetx,offsety     : integer);
```

Input parameters:

Call:	Edit_text(w,col,row,nc,nr,t,x0,y0);
w:	Current edit window
col, row:	Position of top-left corner of sub-window within edit window
nc, nr:	Number of columns and rows in sub-window
t:	Pointer to edited text
x0, y0:	Initial position for top-left corner of sub-window within the text

The routine `Edit_text` can be used to edit one-line and multi-line text referenced by a text pointer. If the parameter $nr = 1$, then the text will be restricted to one line. Otherwise, text of any length can be generated through the sub-window. In the former case, Enter terminates the edit session. In the latter case, the cursor simply moves to the next line of text. In both cases pressing Tab → or Tab ← successfully terminates the edit. In both cases pressing Esc or Break terminates the edit leaving the text unchanged.

The parameters x0 and y0 indicate where in the text editing is to start. They point to the position of the top left of the sub-window.

Routine name:	View text
Kept in unit:	Cit_edit
Purpose:	View a text object

Signature:

```
PROCEDURE View_text(w                      : Citwindow;
                    col,row,ncols,nrows : integer;
                    t                      : Textptr;
                    offsetx,offsety       : integer);
```

Input parameters:

Call: `View_text(w,col,row,nc,nr,t,x0,y0);`
               Parameters as for `edit_text`

The cursor keys and the function keys F5 , F6 , F7 and F8 will control the cursor. Any of the keys Tab → , Enter , Esc or Break terminate the view session.

⟩ 6.7.2   *Editing expressions*

Expressions are an important text object, constructed as they are from multi-line text. When an expression is edited this text is amended as described above in the description of the routine `Edit_text`. As part of the editing, the expression is also compiled, to check its syntax and use of variable names.

The relevant variable names must be present in the dictionary before a text expression can be compiled. The routine `Declare_variable` should be called for each such variable before calling the routine `Edit_expression`. You will normally call the routine `Undeclare` afterwards to remove the names from the dictionary. Both these routines are included at Level 1 in the unit `Cit_eval` which is described in Chapter 9.

The edit session is normally terminated by pressing $\boxed{\text{Tab} \rightarrow}$ or $\boxed{\text{Tab} \leftarrow}$. $\boxed{\text{Enter}}$ will also end the editing of a one-line expression.

A compilation error is reported directly to the screen using the routine `Report_error` described earlier in this chapter. After an error the edit session is resumed.

Pressing $\boxed{\text{Esc}}$ or $\boxed{\text{Break}}$ terminates the edit directly. The default string is substituted for the parameter `expr`, and there is no test compilation.

Routine name:	`Edit_expression`
Kept in unit:	`Cit_edit`
Purpose:	Edit an expression object

Signature:

```
PROCEDURE Edit_expression(w            : Citwindow;
                          col,row      : integer;
                          ncols,nrows  : integer;
                          prestring    : Linestring;
                          VAR expr     : Textptr);
```

Input parameters:

Call:	`Edit_expression(w,col,row,nc,nr,pre,expr);`
`w`:	Current edit window
`col, row`:	Position of preceding string within edit window
`nc, nr`:	Number of columns and rows in sub-window
`pre`:	String that precedes the sub-window
`expr`:	Pointer to expression text

Output parameters:

`expr`:	Edited text expression

The sub-window containing edited text is positioned with its top left corner at the end of the preceding string.

If the parameter `expr` is set to `nil` initially, then a new text object is allocated in memory.

Routine name:	`Edit_string_expression`
Kept in unit:	`Cit_edit`
Purpose:	Edit an expression string

This routine is similar to the last one, except it operates on an expression string rather than expression text. In all other respects it behaves as above.

Signature:

```
PROCEDURE Edit_string_expression(w          : Citwindow;
                                 col,row,
                                 len         : integer;
                                 prestring : Linestring;
                                 VAR expr  : string);
```

Input parameters:

Call: `Edit_string_expression(w,col,row,`
                                `len,pre,expr);`

w: Current edit window
col, row: Position of preceding string
len: Length of edit field
pre: String preceding the edit field
expr: Default string expression

Output parameters:

expr: Edited string expression

) 6.7.3   *Editing values*

As with the display routines, single values can be of type `Citreal` or `integer`, and in the former case there are three possible formats, *real*, *fixed-point*, and *floating-point*. In all cases there is a field to specify. Within this field the defaults will take on the formats already described in the last section.

The edited strings can be expressions containing constants that have been declared in the dictionary. The special symbols π and ∞ can be invoked by the function keys $\boxed{F3}$ and $\boxed{F4}$.

The edit field is preceded and followed by strings which can contain explanations. If the preceding string is empty, then the edit field is positioned directly at the specified position within the edit window.

Routine name: `Edit_real`
Kept in unit: `Cit_edit`
Purpose: Edit a value displayed in *real* format

Signature:

```
PROCEDURE Edit_real(w            : Citwindow;
                    col,row,field : integer;
                    prestring,
                    poststring    : Linestring;
                    VAR r         : Citreal);
```

Input parameters:

Call:	Edit_real(w,col,row,fld,pre,post,r);
w:	Current edit window
col, row:	Position of preceding string
fld:	Width of edit field
pre, post:	Preceding and following string
r:	Default value

Output parameters:

r:	Edited value

Routine name:	Edit_fixpt
Kept in unit:	Cit_edit
Purpose:	Edit a value displayed in *fixed-point* format

Signature:

```
PROCEDURE Edit_fixpt(w             : Citwindow;
                     col,row,field,
                     ndec          : integer;
                     prestring,
                     poststring    : Linestring;
                     VAR r         : Citreal);
```

Input parameters:

Call:	Edit_fixpt(w,col,row,fld,nd,pre,post,r);
	Parameters as for Edit_real

Output parameters:

r:	Edited value

Routine name:	Edit_float
Kept in unit:	Cit_edit
Purpose:	Edit a value displayed in *floating-point* format

Signature:

```
PROCEDURE Edit_float(w              : Citwindow;
                     col,row,field : integer;
                     prestring,
                     poststring    : Linestring;
                     VAR r         : Citreal);
```

Input parameters:

Call: Edit_float(w,col,row,fld,pre,post,r);
                     Parameters as for Edit_text

Output parameters:

r: Edited value

Routine name:	Edit_integer
Kept in unit:	Cit_edit
Purpose:	Edit an integer value

Signature:

```
PROCEDURE Edit_integer(w              : Citwindow;
                       col,row,field : integer;
                       prestring,
                       poststring    : Linestring;
                       VAR i         : integer);
```

Input parameters:

Call:	Edit_integer(w,col,row,fld,pre,post,i);
w:	Current edit window
col, row:	Position of preceding string
fld:	Width of edit field
pre, post:	Preceding and following string
i:	Default value

Output parameters:

i: Edited value

) 6.7.4   *Editing* arrays

An array of real values can be displayed and edited using one routine call.
In the Toolkit array objects are denoted by variables of type **Arrayptr**.

It is assumed that the array has at least ni rows of columns with n0 real elements, the first nj of which are to be edited. As with the display routines this provides scope to edit row vectors, column vectors and matrices to be displayed.

The array is placed on the screen inside a sub-window which itself is located inside the edit window. The first element of the array will be positioned at the top-left corner of the sub-window. Array elements are separated horizontally by one space.

The array is displayed with one element only highlighted. The highlight can be moved from element to element using the cursor keys. If the array is large, the text is scrolled. When Enter is pressed, then editing is switched on. When finished, pressing the Enter key switches back to the selection stage.

Alternatively, the Tab → and Tab ← keys can be pressed in either the selection or edit stages. In this way it is possible to step through the elements one at a time, moving along the rows and down the columns.

Pressing Esc terminates the overall edit session.

If the Boolean parameter **border** is set true, then a count is displayed of the rows and columns in a border within the sub-window. The border is modified as the array text is scrolled.

There are three routines, giving the option of *real*, *fixed-point* or *floating-point* formats in the display. They all have similar parameter sets.

Routine name:	Edit_array_real
Kept in unit:	Cit_edit
Purpose:	Edit an array of values displayed in *real* format

Signature:

```
PROCEDURE Edit_array_real(w          : Citwindow;
                          col,row     : integer;
                          ncols,nrows : integer;
                          border      : boolean;
                          field       : integer;
                          a           : Arrayptr;
                          ni,nj,n0    : integer);
```

Input parameters:

Call:	`Edit_array_real(w,col,row,nc,nr,`
	`field,bord,@A,ni,nj,n0);`
`w`:	Current edit window
`col, row`:	Position of top-left corner of sub-window
`nc, nr`:	Number of columns and rows in sub-window
`field`:	Width of field for elements
`bord`:	Border flag (set `true` for enumeration)
`@A`:	Pointer to array A
`ni, nj`:	Number of rows and columns edited
`n0`:	Row dimension of array A

Routine name:	`Edit_array_fixpt`
Kept in unit:	`Cit_edit`
Purpose:	Edit an array of values displayed in *fixed-point* format

Signature:

```
PROCEDURE Edit_array_fixpt(w          : Citwindow;
                           col,row    : integer;
                           ncols,nrows : integer;
                           field,ndec : integer;
                           a          : Arrayptr;
                           ni,nj,n0   : integer);
```

Input parameters:

Call:	`Edit_array_fixpt(w,col,row,nc,nr,`
	`field,bord,nd,@A,ni,nj,n0);`
`nd`:	Number of decimal places
	Other parameters as for `Edit_array_real`

Routine name:	`Edit_array_float`
Kept in unit:	`Cit_edit`
Purpose:	Edit an array of values displayed in *floating-point* format

Signature:

```
PROCEDURE Edit_array_float(w          : Citwindow;
                           col,row    : integer;
                           ncols,nrows : integer;
                           field      : integer;
                           a          : Arrayptr;
                           ni,nj,n0   : integer);
```

Input parameters:

Call: `Edit_array_float(w,col,row,nc,nr,`
                                            `field,bord,@A,ni,nj,nO);`
                  Parameters as for `Edit_array_real`

There are three companion routines which allow the arrays to be 'viewed' without the elements being edited. The array of values is displayed within the edit subarea, and the cursor keys provide facilities for scrolling through the text. The only difference is that the characters cannot be changed.

Routine name: `View_array_real`
Kept in unit: `Cit_edit`
Purpose: View an array of values in *real* format

Signature:

```
PROCEDURE View_array_real(w            : Citwindow;
                          col,row      : integer;
                          ncols,nrows  : integer;
                          border       : boolean;
                          field        : integer;
                          a            : Arrayptr;
                          ni,nj,nO     : integer);
```

Input parameters:

Call: `View_array_real(w,col,row,nc,nr,`
                                            `field,bord,@A,ni,nj,nO);`
                  Parameters as for `Edit_array_real`

Routine name: `View_array_fixpt`
Kept in unit: `Cit_edit`
Purpose: View an array of values displayed
                        in *fixed-point* format

Signature:

```
PROCEDURE View_array_fixpt(w            : Citwindow;
                           col,row      : integer;
                           ncols,nrows  : integer;
                           field,ndec   : integer;
                           a            : Arrayptr;
                           ni,nj,nO     : integer);
```

Input parameters:

Call: `View_array_fixpt(w,col,row,nc,nr,`
                                     `field,bord,nd,@A,ni,nj,n0);`
nd: Number of decimal places
                        Other parameters as for `Edit_array_real`

Routine name: `View_array_float`
Kept in unit: `Cit_edit`
Purpose: View an array of values displayed
                        in *floating-point* format

Signature:

```
PROCEDURE View_array_float(w            : Citwindow;
                           col,row      : integer;
                           ncols,nrows  : integer;
                           field        : integer;
                           a            : Arrayptr;
                           ni,nj,n0     : integer);
```

Input parameters:

Call: `View_array_float(w,col,row,nc,nr,`
                                     `field,bord,@A,ni,nj,n0);`
                        Parameters as for `Edit_array_real`

⟩ **6.7.5   Edit range**

Ranges can be displayed, and also edited. We now consider editing the
lower and upper bounds of a range of values.

The routine `Edit_range` complements the routine `Display_range`
described in the last section. There are strings that can be included
before, between and after the value. As seen in the example of Section 6.6.3 these strings can be used to make up a complete statement on
the screen.

The individual values are edited with *real* format.

Routine name: `Edit_range`
Kept in unit: `Cit_edit`
Purpose: Edit lower and upper bounds

Signature:

```
PROCEDURE Edit_range(w            : Citwindow;
                     col,row,field : integer;
                     prestring    : Linestring;
                     midstring    : Linestring;
                     poststring   : Linestring;
                     VAR l,h      : Citreal);
```

Input parameters:

Call: Edit_range(w,col,row,field,
                             pre,mid,post,l,h);

w: Current edit window
col, row: Position of preceding string
field: Width of value fields
pre, mid, post: Separating strings
l, h: Default values of lower and upper bounds

Output parameters:

l, h: Edited values of bounds

Pressing $\boxed{\text{Tab} \rightarrow}$ or $\boxed{\text{Enter}}$ terminates editing each of the fields. Pressing $\boxed{\text{Tab} \leftarrow}$ switches back to the first field. Otherwise, pressing $\boxed{\text{Esc}}$ or $\boxed{\text{Break}}$ aborts the edit, and the default values are returned.

〉*6.7.6   Edit name*

This routine provides the facility for editing a name.

Routine name: **Edit_name**
Kept in unit: **Cit_edit**
Purpose: Edit a name string

Signature:

```
PROCEDURE Edit_name(w             : Citwindow;
                    col,row,field : integer;
                    prestring     : Linestring;
                    poststring    : Linestring;
                    VAR str       : Linestring);
```

Input parameters:

Call:	`Edit_name(w,col,row,field,`
	`            pre,post,s);`
w:	Current edit window
col, row:	Position of preceding string
field:	Width of edit field
pre, post:	Preceding and following string
s:	Default string

Output parameters:

s:	Edited string

Pressing Tab →, Tab ← or Enter terminates the edit; pressing Esc or Break aborts it. In the latter case the parameter **s** is set to the default string.

Any short string can be edited using this routine.

) Chapter 7

) Graphics Routines

) 7.1   Introduction to Level 1

This is the first chapter in the section of the *Scientific Programmer's Toolkit* which deals with Level 1 routines, and so here we recapitulate their purpose briefly: Level 1 routines are component routines out of which higher-level routines can be constructed and which give you a high degree of control over the details of the effect achieved. Level 1 routines themselves use the primitive routines from the unit `Cit_prim`.

Level 1 routines for graphics are described in this chapter. The remaining Level 1 topics (screen and text utilities, and routines for compiling and evaluating expressions) are described in Chapter 8. Some details of certain primitive routines which may commonly be required by the Level 1 user are also included in each chapter as appropriate.

) 7.2   How errors are dealt with

Section 3.2 discusses the principles of error handling for Level 2 routines. You will recall that there are three global variables which are used for holding information about errors. These variables and their types are:

- `Errorflag`   : Boolean;
- `Errorcode`   : Integer;
- `Errorstring` : Linestring;

Once `Errorflag` has been set `True` (i.e. an error has been detected), most Level 2 routines exit immediately. This is not appropriate at Level 1 since the purpose of Level 1 is to give the programmer more detailed control, and it is now the programmer's responsibility not to call a routine under inappropriate circumstances. The rule at Level 1 is therefore that the state of the global error variables does not affect the action of Level 1 routines, and further that the settings of the global error variables are preserved unless a new error condition is met. If an error is detected, then the error globals are reset to the new error and the old settings are lost.

⟩ 7.3   Introduction to Level 1 graphics routines

Unit: `Cit_grap`

The *Scientific Programmer's Toolkit* is designed for portability. That is, we have designed the Toolkit assuming only that certain very elementary screen operations are available, the relevant ones for this chapter being for example the facility to join any two pixels (points on the screen) with a line, draw a solid filled rectangle, *etc.* It is the job of the primitive routines to provide a standard interface to such facilities, so that although the primitives themselves will have to be re-written for different machines, the higher-level routines that use the primitives should be able to work without change. It is therefore desirable that the primitives should be as simple as possible, and they therefore 'talk' to the screen in screen units. The main job of the Level 1 graphics routines is to enable you to 'talk' to the screen in problem units, and to provide other basic facilities of a general nature that can be built out of the primitives. We will now summarise these facilities under three categories, followed by a section on related primitive routines (such as routines to control the current graphics and text colours).

⟩ 7.3.1   The graph port category

All graphics drawing is done within a `Graph_port`, which is simply a rectangular area of the screen defined in terms of screen coordinates. The routines that do the drawing will clip all lines and shapes so that they lie within the port. There are two routines to set this port up: `Graph_port` defines the port in terms of text coordinates, and `Graph_window` defines the port in terms of a previously defined graphics window. (See Chapter 6 for details of text coordinates and graphics windows.) It is also possible to set up a port in terms of screen coordinates using a Level 0 routine `Gp_viewport`, but in our experience this is rarely necessary. As one frequently wishes to mark out the graphics port, there is a routine to draw a border round its extremity.

⟩ 7.3.2   *The scaling and axes category*

Defining all the details needed to calculate scale factors to convert user-coordinates to screen coordinates, draw and label axes, *etc.*, is surprisingly complicated. So many items of data are required that it would be quite unwieldy to set them by passing them as arguments in procedure calls. Instead, we use *graphics parameters*. A full description of these (and there are over 40 of them) will be given later in the chapter; they include items such as the user-coordinate values assigned to the ends of each axis, the number of intervals to be marked off along each axis, the size of the axis tick marks, *etc.*, *etc.* A significant advantage of using parameters is that they are all given default values automatically, and we have designed them in such a way that if you leave the defaults unchanged, apart from a few really vital ones like the axis ranges, you will usually get some kind of intelligible result which you can fine-tune later if you wish.

There are routines to reset the system defaults for the graphics parameters, to save them to a data structure and reload them later, to set scale factors on the basis of the parameters, to draw axes, to label them, to rotate the axes, to draw a coordinate grid, and to convert user-coordinates into screen coordinates.

This is a good place to point out that all the axes and scaling operations are done in 3-D. However the default parameters set up a viewpoint directly above the z-axis so that graphs will appear as normal 2-D x–y plots unless you change the viewpoint. This greatly simplifies the use of 3-D graphics without in any way complicating matters for 2-D graphs.

⟩ 7.3.3   *The plotting category*

Once user-axes have been defined together with the graph port, then plotting can be done. There are routines to move to points defined by two coordinates (x, y), or three (x, y, z), and to draw lines in a selected colour to other points. Movement can be absolute or relative to the last point. There are routines for labelling graphs, plotting symbols, using cross-wires, and drawing polygons either in outline or filled.

⟩ 7.3.4   *Related primitives*

When using the plotting routines, there are three characteristics which will affect the resulting plots: *colour*, *style* and *writemode* (sometimes also called *transfer mode*). The effect of colour needs no further description (but refer to Chapter 1 for a description of how the Toolkit handles the colour capabilities of different graphics hardware). By style is meant, for example, whether a line is solid, dotted, dashed, *etc*, or for an area what kind of pattern is used to fill it. By transfer mode is meant how the object being plotted interacts with whatever is already on the screen: for example, replacing the screen pixels with the object's pixels, 'OR'ing the colours for coincident pixels, *etc*.

Unlike in the Level 2 routines, these characteristics are not included in the routine arguments because it is often necessary to call the plotting routines repeatedly as, for example, when drawing a curve out of many line segments. Instead, there are current settings for each of these characteristics, which are controlled by primitive level routines. The reason they are classed as primitives is that they are fundamental to any screen operation, not just to those which relate to mathematical graph plotting.

⟩ 7.4   Graph port routines

The **graph port** is a rectangular area of the screen within which graphics operations are done. There is a primitive routine **Gp_viewport** (documented in with other Level 0 routines) which allows you to specify the graph port in terms of screen coordinates, but as these will vary between different graphics hardware, it is usually more convenient to use one of the Level 1 routines.

If you attempt to draw a line or any other graphics construction so that all or part of it lies outside the port, then it is clipped or masked so that only the portion within the port will show. Data on the graph port are used when setting up axes and scale factors.

You should call either **Graph_port** or **Graph_window** before calling any other graphics routine. These calls, as well as defining the port, will ensure that the graphics hardware is in graphics mode. (There is also a primitive routine **Graphics_mode** which you can use if you want to avoid using the port routines for any reason.)

Signature:

```
PROCEDURE Graph_port(x1,y1,x2,y2: integer);
```

Graph_port defines the port in terms of text coordinates, which are described fully in Chapter 8, but which can easily be understood with the help of figure 7.1. **x** text coordinates run horizontally left to right with the leftmost character position counted as 1, so the rightmost position is equal to the number of characters per line on the screen. **y** text coordinates run vertically from the top down with the top line counting as 1, so the bottom line has a number equal to the number of lines to the screen. Coordinates are presented to the routines in the order **x1** (least **x**), **y1** (least **y**), **x2** (greatest **x**), **y2** (greatest **y**). You can think of these as the coordinates of two cells, top-left and bottom-right. Note that the cell referenced is included in the graph port.

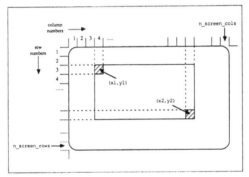

Figure 7.1 Graph ports and text coordinates

Signature:

```
PROCEDURE Graph_window(w: Citwindow);
```

Graph_window defines the port in terms of a previously defined graphics window (see Chapter 6). Windows are associated with a record data structure which contains an array **wint[1..4]** giving the internal coordinates of the window in screen units. These are simply copied to the graph port. (You do not need to remember the details. The values are calculated and set when you define a window: the information is mentioned here for completeness.) The window record also contains three colour pairs. The first colour of the first pair **wnd.txtcolor** is used to set the colours for labelling axes, and the third pair **wnd.auxcolor1** and **wnd.auxcolor2** are used to set the pip and axis colour, respectively.

These settings are copied to the corresponding graphics parameters (see next section) during the call to **Graph_window**, so you can easily override any setting by adjusting the parameters. Recall too that current colours, style and transfer mode can be changed using one of the primitive routines described later.

Graph_window is the most convenient way to set up graphics windows if you want to make several changes to defaults. Using it ensures compatibility with any other windows which you use. The examples **CHAP3\E-3DRAW** and **CHAP7\3-ANGLES** show how it is used.

Signature:

```
PROCEDURE Draw_border;
```

This routine simply draws the border at extremity of port, using the current graphics colour.

⟩ 7.5   Graphics parameters

There are two kinds of graphics parameters: *primary parameters* and *secondary parameters*. The secondary parameters are private to the unit **Cit_grap** and we will not be concerned further with them except to note that they consist of quantities calculated from the primary parameters. The calculations are done when the routine **Set_scales** is invoked. You should therefore ensure that the primary parameters are set as you want before calling **Set_scales**, and you should not alter them until your current plotting activities are complete and you are ready to go on to another plot. We shall refer to the primary graphics parameters simply as graphics parameters in the rest of this chapter.

Graphics parameters are stored in a Pascal record which is declared as a **TYPE** in unit **Cit_grap** as shown below. A variable called **Graphparam** of this type is also declared. You have access to both the type and the variable in your part of the program provided that you include **Cit_grap** in your **USES** statement.

Declaration of graphics parameters:

```
TYPE Citgraphprimaries = record
   xlo, ylo, zlo        : Citreal {axis ranges};
   xhi, yhi, zhi        : Citreal;
   xorig, yorig, zorig  : Citreal {axes crossing};
   th, ph               : Citreal {view angles};
   zviewplane           : Citreal {2d view plane};
   nxi, nyi, nzi        : integer {no. of intvls};
   wlf, wbt, wrt, wtp   : integer {window position};
   mlf, mrt, mbt, mtp   : integer {window margins};

   xdisp, ydisp, zdisp
                : integer {label displacement};
   labeldx, labeldy
                : integer {graph label displacement};
   xfield, yfield, zfield
                : integer {label field size};
   xpipm, ypipm, zpipm : byte    {pip mode};
   xpipl, ypipl, zpipl : byte    {pip length};
   xflag, yflag, zflag : byte    {axis flags };
   caxis, cpip, clab
                : Citcolor {axis, pip & label colors};
   cline        : Citcolor {default line colour}
   end;
```

{ Primary parameters for current graph }
VAR Graphparam: Citgraphprimaries;

You can refer to any parameter with the usual Pascal construction for fields within records. For example,

`Graphparam.xlo := -10`

will reset **xlo**. If you want to reset several parameters you can use the **with** construction. For example:

```
with Graphparam do
   begin
   xlo:=-5; xhi:=5;
   ylo:=-2; yhi:=2
   end:
```

In the rest of this chapter we shall omit the record name and refer to **Graphparam.zlo** as just **zlo**, *etc.*

⟩ *7.5.1   Axis ranges*

Parameters: **xlo, ylo, zlo, xhi, yhi, zhi**
     Defaults: -10, -10, -10, 10, 10, 10

These are the values in user-units which the user wants to be included
in the range of each axis. They are not necessarily the values of the axis
ends as actually drawn, since the Toolkit will round the range up to be
a multiple of a sensible interval size.

⟩ *7.5.2   Origin*

Parameters: **xorig, yorig, zorig**
     Defaults: 0,   0,   0

These allow you to control the coordinates of the point where the axes
cross. If one of these values lies outside the corresponding axis range,
then the Toolkit will select a crossing point automatically.

⟩ *7.5.3   View angles and plane*

Parameters: **th,   ph,   zviewplane**
     Defaults: 0, $-\frac{\pi}{2}$,   0

Toolkit graphics is all done with 3-D coordinates projected onto the
screen to give a perspective view, and the angles from which you view
are **th** and **ph** (in radians) relative to the 3-D axes. Mathematically
these angles are commonly known as θ (*theta*) and ϕ (*phi*). They are
shown in figure 7.2. If OA is the line of sight, then θ is the angle between
OA and OZ (the z-axis), and ϕ is the angle between the plane OAZ and
the x-axis, measured counter-clockwise. You can see an animation of
the effect of varying θ and ϕ in example **3-ANGLES** described later in the
chapter.
     2-D graphics is simply a special case of 3-D. By taking $\theta = 0$ you look
down on the x–y plane from directly over the z-axis which therefore
appears only as a dot. The angle ϕ now controls the rotation of the
axes: $\phi = -\frac{\pi}{2}$ gives the usual x–y configuration, which is the default.
     Since 2-D plotting is so common it would be tedious to have to specify
three coordinates (x, y, z) every time a graphics point was required, and
many routines are specific to 2-D, **Draw_curve1** for example in Chapter 3. The parameter **zviewplane** is used in these cases when the Toolkit

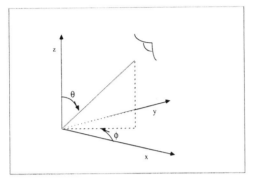

Figure 7.2 3-D viewing angles θ and ϕ

routines are calculating screen coordinates. It is rarely necessary to change the default.

> 7.5.4   *Axis intervals*

Parameters: **nxi,   nyi,   nzi**
  Defaults: 7,   7,   7

These integer parameters are used to determine the interval sizes along each axis. They will not necessarily be the actual number of intervals since the routines adjust the interval size to a standard form which is 1, 2 or 5 times a power of 10.

> 7.5.5   *Window data*

Parameters: **wlf,   wtp,   wrt,   wbt**
  Defaults: 0,   0,   *,   *
Parameters: **mlf,   mtp,   mrt,   mbt**
  Defaults: *,   *,   *,   *

These parameters are copies of the current graphics window (port) data. They are adjusted whenever one of the graph port routines are called (described above). The default values marked * depend on the screen hardware and mode being used and give the full screen with a one character width margin at the left and right and a one-and-a-half character height margin at top and bottom.

⟩ 7.5.6   *Label position and control*

Parameters: **xdisp, ydisp, zdisp**
    Defaults: *,   1,   0
Parameters: **labeldx, labeldy**
    Defaults: 0,   *
Parameters: **xfield, yfield, zfield**
    Defaults: 5,   5,   5

There are two kinds of labelling: the automatic labelling of axes with the values corresponding to the axis ends and pips, and labelling with text strings at the user's specific request. In both cases it is convenient to apply a standard displacement to the screen graphics coordinates at which this text will be printed.

xdisp, ydisp, zdisp, and **xfield, yfield, zfield** are used only by the **Label_axes** routine (described below). The **xfield** *etc.* parameters determine the number of character spaces that will be used when converting the axis label variable values into strings using the **Format_real** routine (see Chapter 8). **xdisp** *etc.* are in screen coordinates. The default for **xdisp** is $(1 - \text{char\_ht}/2)$.

labeldx and **labeldy** are used in conjunction with **Graphlabel** and **Graphlabel3** (see below). The default value for **labeldy** is $\text{char\_ht}/2$.

⟩ 7.5.7   *Control of the axis pips*

Parameters: **xpipm, ypipm, zpipm**
    Defaults: 2,   2,   2
Parameters: **xpipl, ypipl, zpipl**
    Defaults: 2,   2,   2

The 'pips' on each axis can be drawn in one of three modes, selected independently for each axis according to the value of **xpipm** *etc.* Mode 2 gives a pip centred on the axis, mode 1 makes one end of the pip start from the axis and project upwards or rightwards, and mode 0 is like mode 1 except that the pip projects downwards or leftwards.

xpipl *etc.* controls the pip length. The values of **xpipl** *etc.* (integers) are taken as percentages and applied to the length of the axis running at right angles to the axis being 'pipped'.

⟩ *7.5.8   Axis flags*

Parameters: **xflag, yflag, zflag**
   Defaults: 1,   1,   0

These flags determine whether a particular axis will be drawn or labelled. When either of the routines **Draw_axes** or **Label_axes** are called, if a flag is zero then the operation on the corresponding axis is skipped. The flag **zflag** is set to zero for 2-D plotting.

⟩ *7.5.9   Axis colours and default line colour*

Parameters: **caxis,   cpip,   clab,   cline**
   Defaults: **AXISRED, PIPYELLOW, LABLIGHTGRAY, LINEWHITE**

These parameters are of type **Citcolor** and determine the colour of axes, pips, labels and the default for graph drawing respectively. Parameters **caxis** and **cpip** take effect only during calls to **Draw_axes**, and **clab** takes effect only in **Label_axes**, but at the end of each of these routines the graph colour is set to **cline** which will therefore be the graphics colour for subsequent drawing unless changed deliberately by the user (by using **Gp_graph_color** for example). Note also that **cline** is set to the graphics colour associated with the selected window when **Graph_window** is called, so that it should never be necessary to set **cline** directly.

   This is an appropriate moment to mention the special colours which are used as defaults. They are defined in the public interface of unit **Cit_grap** and may be referenced in your own program if you have included this unit in your **USES** statement:

```
{ Default graph colours }
CONST AXISRED       : Citcolor = (1,2,4,4);
PIPYELLOW           : Citcolor = (1,3,14,14);
LABLIGHTGRAY        : Citcolor = (1,3,7,7);
```

   The difference between these colours and the standard colours **CITRED**, *etc.* lies in the colour substitution that is done when working in graphics modes with less than 16 colours. In the standard colours, red is thought of as dark and so is mapped to other dark colours, and ultimately to black in two-colour modes (monochrome). So if you used **CITRED** for example for the axis colour, it would not be visible if you switched to monochrome. These three special colours are always visible against the standard background. If this is not the effect you want, you can alter the default.

⟩ 7.6   Graphics parameter procedures

There are four procedures which deal with the graphics parameters as a body. They are designed to help you switch between two or more graphics environments, or to set up your own files of standard parameter settings.

Signature:

```
PROCEDURE Default_primaries;
```

This procedure resets the primary graphics parameters to their default values.

Signatures:

```
TYPE Paramptr = ^char;
PROCEDURE Save_graph_param(var p: Paramptr);
PROCEDURE Load_graph_param(p    : Paramptr);
PROCEDURE Free_graph_param(var p: Paramptr);
```

The type **Paramptr** is used to reference heap parameters. The routine **Save_graph_param** copies the values of the primary and secondary parameters to space allocated on the heap, and sets the pointer **p**. **Load_graph_param** restores all the primary and secondary parameters to the values in the store referenced by **p**.

You should not call **Save_graph_param** twice with the same argument p without an intervening call to **Free_graph_param**, otherwise you will leave some 'dead' space on the heap.

Examples **3-ANGLES** and **4-TWIN** demonstrate the use of these routines.

⟩ 7.7   Scaling and axes routines

⟩ 7.7.1   *Standard routines*

Signatures:

```
PROCEDURE Set_scales;
PROCEDURE Draw_axes;
PROCEDURE Label_axes;
```

Having set the primary graphics parameters as required, the call **Set_scales** calculates the secondary parameters (scale factors, *etc.*) so that the plotting commands can be used. It is therefore *essential* to call **Set_scales**. However you need not use **Draw_axes** or **Label_axes** if you do not want the axes drawn or labelled. If you only want some of the axes drawn or labelled, set the parameters **xflag** *etc.* to 0 for the axes you want suppressed.

⟩ 7.7.2   *Additional routines*

Signatures:

```
PROCEDURE Draw_box;
PROCEDURE Draw_grid;
PROCEDURE Rotate_axes(dph,dth: Citreal);
```

Draw_box and **Draw_grid** do what they say. **Rotate_axes** adds **dph** to **ph** (ϕ), and **dth** to **th** (θ) and recalculates other parameters affected without changing the scale factors or the axes crossing point. The rotated axes are not drawn or labelled: **Draw_axes** and **Label_axes** should be used if required. Example **2-POLY** below includes a call to **Rotate_axes**.

⟩ **7.8   Conversion to screen coordinates**

Signatures:

```
FUNCTION Convert_x (x,y  : Citreal): integer;
FUNCTION Convert_y (x,y  : Citreal): integer;
FUNCTION Convert_x3(x,y,z: Citreal): integer;
FUNCTION Convert_y3(x,y,z: Citreal): integer;
```

Once scales *etc.* have been set the transformation from user-coordinates to screen coordinates is defined. Normally you will not need to know the screen coordinates of the points which you plot; these conversion routines are provided for completeness.

Convert_x and **Convert_y** are for use in the 2-D case. They are equivalent to the calls
```
Convert_x3(x,y,zviewplane)
Convert_y3(y,x,zviewplane)
```
respectively, where the second two arguments are parameters.

⟩ **7.9   Moving and drawing in user-coordinates**

Signatures:

```
PROCEDURE Moveabs (x,y  : Citreal);
PROCEDURE Moverel (x,y  : Citreal);
PROCEDURE Moveabs3(x,y,z: Citreal);
PROCEDURE Moverel3(x,y,z: Citreal);
PROCEDURE Drawabs (x,y  : Citreal);
PROCEDURE Drawrel (x,y  : Citreal);
PROCEDURE Drawabs3(x,y,z: Citreal);
PROCEDURE Drawrel3(x,y,z: Citreal);
```

These routines provide basic moving and drawing using user-coordinates. The drawing routines use the current line colour and writemode, which may be changed using **Gp_graph_color** and **Gp_writemode**. By default this is 'actual mode'. See the appendices for details of available colours (Section A1.5) and writemodes (Section A2.3). The 2-D routines are equivalent to the 3-D versions with **z** replaced by **zviewplane**.

〉 7.10   Labels, symbols and cross-wires

Signatures:

```
PROCEDURE Graphlabel
(x,y  : Citreal; s      : Linestring);
PROCEDURE Graphlabel3
(x,y,z: Citreal; s      : Linestring);
PROCEDURE Plot_symbol
(x,y  : Citreal; symbol: integer);
PROCEDURE Plot_symbol3
(x,y,z: Citreal; symbol: integer);
PROCEDURE Cross_wires (x,y  : Citreal);
PROCEDURE Cross_wires3(x,y,z: Citreal);
```

The first pair of these allow you to write text strings in the current text colour at a point specified in user-coordinates. See the description of **labeldx** and **labeldy** in Section 7.5 above.

The next pair plot symbols (. × or +) at the point specified. **Symbol** is 0, 1 or 2, respectively.

The cross-wires routines generate two or three lines parallel to the axes which intersect at the specified point.

〉 7.11   Polygons

Signatures:

```
PROCEDURE Draw_polygon (x,y  : Arrayptr;
                        n,n0 : integer);
PROCEDURE Fill_polygon (x,y  : Arrayptr;
                        n,n0 : integer);
PROCEDURE Draw_polygon3(x,y  : Arrayptr;
                        n,n0 : integer);
PROCEDURE Fill_polygon3(x,y  : Arrayptr;
                        n,n0 : integer);
```

The draw routines generate outlines, and the fill routines generate filled shapes. Both use the current graphics colour. **n** is the number of points defining the polygon, and the coordinates (in user-units) are taken from the arrays referenced by **x** and **y**. Suppose that successive vertices of the polygon are (x_1, y_1), (x_2, y_2), ... *etc*. Then the values $x_1, x_2, ...$ must be stored in the array referenced by **x**, and the values $y_1, y_2, ...$ must be stored in the array referenced by **y**. For three-dimensional polygons an extra array **z** is used for the z-values.

Coordinate values may be stored either as elements of a one-dimensional array or as columns of a two-dimensional array. (This technique is discussed with examples in Section 3.6.) In the one-dimensional case, **n0** should be set to 1 and in the two-dimensional case **n0** is the row dimension of the array (see **poly2** in the example below for which **n0** = 2).

```
TYPE
  polydata1 = ARRAY[1..4] OF Citreal;
  polydata2 = ARRAY[1..4,1..2] OF Citreal;
CONST
  xdata : polydata1=( 2, 2, -1, -1);
  ydata : polydata1=(-1, 1,  1, -1);
  poly2 : polydata2=((2,-1),(2,1),(-1,1),(-1,-1));
  ...
Draw_polygon(@xdata, @ydata, 4,1);
Draw_polygon(@poly2, @poly2[1,2], 4,2);
```

If the two-dimensional array is used, then the row dimension must be 2 for two-dimensional graphs and 3 for three-dimensional graphs. The example 2-POLY shows **fill_polygon** in use.

) 7.12   Related primitives

Signatures:

```
PROCEDURE Gp_graph_color (clr : Citcolor);
PROCEDURE Gp_line_style  (ibmstyle : integer);
PROCEDURE Gp_writemode   (writemode : byte);
```

Section 7.3.4 has introduced the purpose of those primitives which are related to this unit.

The structure of **Citcolor** types has been explained in Chapter 1, and a list of predefined colours is given in Section A1.5. The argument **clr** for **Gp_graph_color** can be a predefined colour or one which you have defined yourself.

The following constants (defined in unit CIT_PRIM) are available for line styles: solid_line, dashed_line, and dotted_line. The default is solid_line. The names themselves describe the effect.

The following constants (defined in unit CIT_PRIM) are available for writing modes (also known as transfer modes): actual_mode, invert_mode, eor_mode, or_mode and and_mode. The default is actual_mode. Because of the colour mapping which the Toolkit does, the effect of modes other than actual mode is difficult to predict precisely. If drawing a line in actual mode, for example, then the line is drawn on top of anything else that may be on the screen and all pixels on the line are changed to the current graphics colour, which will be specified by the current graphics colour according to how many screen colours are available. If the same line were to be drawn using invert mode, the current graphics colour is irrelevant as the colours of all the pixels on the line are inverted: exactly what the resulting colour will be depends on the hardware. However, drawing the same line again will restore the original colours. In 'eor mode' ('exclusive or'), the binary value of the colour associated with each pixel is exclusive-or'ed with the current graphics colour. As with inversion a repeat operation, assuming no intervening changes to the affected part of the screen, restores the original image. Finally, 'or mode' and 'and mode' combine colours using the logical 'or' or 'and' operations which results in a combination of the object being drawn with the existing screen image.

⟩ **7.13   Examples**

⟩ *7.13.1   Basic use of Level 1 graphics*

Example 1-GRAPH shows the most basic use of Level 1 graphics. Note the order of operations: modify the graphics parameters, set the scales, draw axes, label axes, move to the first point, then loop move to successive points until the graph is complete.

```
{ 1-graph            RDH 4/5/89 }
{ First Level 1 example: basic use }

PROGRAM Graph1;
```

```
    USES
      Cit_core, Cit_prim, Cit_grap;

    CONST
      N = 32;

    VAR
      i        : integer;
      x, xstep : Citreal;

    FUNCTION f (x : Citreal) : Citreal;
    BEGIN
      f := x * (x * x - 2)
    END;

BEGIN

  Graphics_mode;

  {set up axes for x=(-2,2) y=(-4,4)}
  WITH Graphparam DO BEGIN
    xlo := - 2;
    xhi := 2;
    ylo := - 4;
    yhi := 4
  END;

  Set_scales;
  Draw_axes;
  Label_axes;

  xstep := (Graphparam.xhi - Graphparam.xlo) / N;
  x := Graphparam.xlo;

  Moveabs (x, f (x));
  FOR i := 1 TO N DO BEGIN
    x := x + xstep;
    Drawabs (x, f (x))
  END;

END.
```

) *7.13.2 Changing the defaults*

Example 2-POLY is a little more adventurous and shows several plots
using different graph ports with the axes being rotated, colour changes,
and use of fill_polygon.

```
{ 2-poly                        RDH 14/9/89 }
{ Second Level 1 example: several ports, etc }


PROGRAM Poly2;

  USES
    Cit_core, Cit_prim, Cit_grap;

  CONST
    N_DATA = 4;
  TYPE
    poly_data = ARRAY [1..N_DATA,
      1..2] OF Citreal;
  CONST
    POLY:poly_data = ((2, - 1), (2, 1), (- 1, 1),
      (- 1, - 1));

  PROCEDURE our_polygon (txt : Linestring;
                         angle : Citreal;
                         clr : Citcolor);
  BEGIN
    Graphparam.th := 0.0;
    Set_scales;
    Rotate_axes (angle, 0);
    Draw_axes;
    Draw_box;
    Graphlabel (- 1.5, - 1.5, txt);
    Gp_graph_color (clr);
    Fill_polygon (@POLY, @POLY[1,2], N_DATA, 2)
  END;
```

```
BEGIN

  {set up axes for x=(-2,2) y=(-4,4)}
  WITH Graphparam DO BEGIN
    xlo := - 3;
    xhi := 3;
    ylo := - 3;
    yhi := 3
  END;

  Graph_port (14, 2, 38, 12);
  our_polygon ('A', 0.0, CITWHITE);

  Graph_port (40, 2, 64, 12);
  our_polygon ('B', Pi / 3, CITGREEN);

  Graph_port (14, 14, 38, 24);
  our_polygon ('C', 2 * Pi / 3, CITBROWN);

  Graph_port (40, 14, 64, 24);
  our_polygon ('D', Pi, CITYELLOW);

END.
```

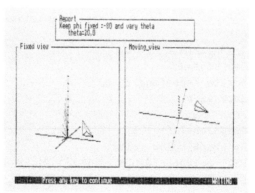

Figure 7.3 3-D viewing angles demonstrated by program **3-ANGLES**

) 7.13.3 A 3-D example

Example 3-ANGLES uses text windows to define the graphics windows.
Level 1 routines are used to set up axes and scales, and then an object is
drawn in two windows. In the first, the viewing angles are kept constant
and a succession of view vectors are drawn. For each view, the object is
redrawn in the second position as it would appear from the latest view
direction.

```
{ 3-angles                    RDH 14/9/89 }
{ Level 1 example: 3-D viewing angles etc  }


PROGRAM Angles;

  USES
    Cit_core, Cit_prim, Cit_wind, Cit_grap,
      Cit_ctrl, Cit_disp;

  CONST
    N_DATA = 4;
  TYPE
    object_data = ARRAY [1..N_DATA,
      1..3] OF Citreal;
  CONST
    OBJECT:object_data = ((2, 1, 0), (1, 1, 0),
      (1, 2, 0), (1, 1, 1));

  VAR
    text_w, fixed_w, view_w : Citwindow;
    fixed_params : Paramptr;
    theta, phi : Citreal;
    i : integer;

  {Convert degrees to radians}
  FUNCTION rad (a : Citreal) : Citreal;
  BEGIN
    rad := Pi * a / 180
  END;
```

```
{Draw the edges of tetrahedron}
PROCEDURE outline;
  {join vertices i and j}
  PROCEDURE join (i, j : integer);
  BEGIN
    Moveabs3 (OBJECT [i, 1], OBJECT [i, 2],
      OBJECT [i, 3]);
    Drawabs3 (OBJECT [j, 1], OBJECT [j, 2],
      OBJECT [j, 3])
  END;

BEGIN
  Gp_graph_color (CITGREEN);
  join (1, 2);
  join (1, 3);
  join (1, 4);
  join (2, 3);
  join (2, 4);
  join (3, 4)
END {of outline};

{Draw the outline in view window from new view}
PROCEDURE new_view (theta, phi : Citreal);
  VAR
    r, s : Citreal;
BEGIN
  {convert to radians}
  theta := rad (theta);
  phi := rad (phi);
  {set the new view in moving view window}
  Graph_window (view_w);
  WITH Graphparam DO BEGIN
    th := theta;
    ph := phi
  END;
  {draw the new view}
  Clear_window (view_w);
  Set_scales;
  Draw_axes;
  outline;
```

```
  {restore params for fixed view}
  Load_graph_param (fixed_params);


  {and draw view vector}
  Gp_graph_color (CITWHITE);
  Moveabs3 (0, 0, 0);
  r := 3;
  s := r * Sin (theta);
  Drawabs3 (s * Cos (phi), s * Sin (phi),
    r * Cos (theta))
END;


{ Main program }
BEGIN


  { Set up the windows }
  Define_window (text_w, 15, 2, 65, 5,
    {window position}
    CITYELLOW, CITRED,
    { colours of text   }
    CITWHITE, CITRED,
    { ... of margins }
    CITWHITE, CITBLACK,
    { ... of highlights }
    MARGINS, False);


  Define_window (fixed_w, 1, 6, 38, 23,
    {window position}
    CITLIGHTGRAY, CITBLACK,
    {colours of body}
    CITWHITE, CITBLACK,
    { ... of margins }
    PIPYELLOW, AXISRED,
    { ... of axes }
    MARGINS, False);
```

```
Define_window (view_w, 40, 6, 79, 23,
  {window position}
  CITLIGHTGRAY, CITBLACK,
  {colours of body}
  CITWHITE, CITBLACK,
  { ... of margins }
  PIPYELLOW, AXISRED,
  { ... of axes }
  MARGINS, False);


{ switch on graphics mode ... }
Graph_window (fixed_w);


{ open windows ... }
Open_window (text_w, 'Report');
Open_window (fixed_w, 'Fixed view');
Open_window (view_w, 'Moving_view');


{set up axes and fixed viewing angles}
WITH Graphparam DO BEGIN
  xlo := - 3;
  xhi := 3;
  ylo := - 3;
  yhi := 3;
  zlo := 0;
  zhi := 4;
  th := rad (70);
  ph := - rad (70);
  zflag := True
END;


{Draw the fixed view and save parameters}
Set_scales;
Draw_axes;
outline;
Save_graph_param (fixed_params);
```

```
{Keep theta fixed and vary phi}
theta := 80;
Display_real (text_w, 2, 1, 3,
  'Keep theta fixed =', ' and vary phi', theta);
i := 0;
REPEAT
  phi := i * 18 - 80;
  Display_real (text_w, 5, 2, 4, 'phi=', Blanks,
    phi);
  new_view (theta, phi);
  Pause;
  i := i + 1
UNTIL (i > 20) OR Escape;


{Keep phi fixed and vary theta}
phi := - 80;
Display_real (text_w, 2, 1, 3,
  'Keep phi fixed =', ' and vary theta', phi);
Write_string (text_w, 5, 2, Blanks);
Pause;
{Re-draw the fixed view with saved parameters}
Clear_window (fixed_w);
Load_graph_param (fixed_params);
Set_scales;
Draw_axes;
outline;


i := 0;
REPEAT
  theta := 80 - i * 10;
  Display_real (text_w, 5, 2, 4, 'theta=',
    Blanks, theta);
  new_view (theta, phi);
  Pause;
  i := i + 1
UNTIL (i > 8) OR Escape;


END.
```

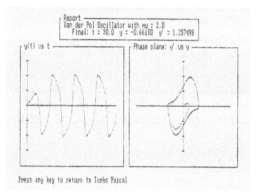

Figure 7.4 The display produced by program **4-TWIN**

) *7.13.4   Simultaneous plotting*

Example **4-TWIN** shows how loading and saving graphics parameters
can be used to build up two graphs at once, and also demonstrates a
combination of mathematical and graphics routines. It solves the same
problem (the differential equation known as the Van der Pol oscillator)
as in Chapter 5, except that the results are plotted as calculated simultaneously in two windows. In one window we plot y against t, and in
the other y' against y (known as a phase-plane plot). It would be easier
programming to plot the first graph completely, and then do the second,
but then the graphs would not be seen to be growing simultaneously.
Notice that when you do 'live' plotting in this way it is not easy to set
up the axes automatically, since the ranges of the variables cannot be
known until the calculation is complete.

```
{4-twin.pas ( based on 5-3vdp2)                          }
{- Van der Pol oscillator using RKF                      }
{    Modified to plot y vs t and y' vs y in twin         }
{    windows. (Phase plane plot.) - SMW/RDH  21/9/89}
```

```
PROGRAM Vdptwin;

  USES
    Cit_core, Cit_prim, Cit_wind, Cit_grap,
      Cit_math, Cit_ctrl, Cit_disp;

  CONST
    MU    = 2.0;
    ORDER = 2;
    DTMIN = 0.0001;
    RELER = 1E-5;
    ABER  = 1E-5;
    NMAX = 500;

  VAR
    t, dt, oldstep : Citreal;
    y, yp : ARRAY [1..ORDER] OF Citreal;
    n, nstepsleft : integer;
    ok : boolean;

    {windows and parameter blocks}
    text_w, yt_w, phase_w : Citwindow;
    yt_pars, phase_pars : Paramptr;
    {to remember last point}
    t_last, y_last, yp_last : Citreal;

  { Example function for the second order }
  { Van der Pol non-linear oscillator. }
  { Note mu = 0 gives harmonic oscillator }
  { y'' + mu * y' * (y ^ 2 - 1) + y = 0 }

{$F+}
  FUNCTION f : Citreal;
  BEGIN
    yp [1] := y [2];
    yp [2] :=
      - y [1] - MU * y [2] * (Sqr (y [1]) - 1.0);
    f := 0
  END { f };
{$F-}
```

```
PROCEDURE initialise;


  CONST
    { stopping value for T }
    TFINAL = 30.0;


BEGIN
  y [1] := 0.1;
  y [2] := 0.1;
  t := 0.0;
  dt := 0.2;
  { set number of steps }
  nstepsleft := Round (TFINAL / dt);
  n := 0;
  { ensure step size divides tfinal }
  dt := TFINAL / nstepsleft
END { initialise };


PROCEDURE print (txt : Linestring);
BEGIN
  Display_real (text_w, 5, 2, 6, txt + ' t = ',
    ' ', t);
  Display_real (text_w, 22, 2, 8, 'y = ', ' ',
    y [1]);
  Display_real (text_w, 36, 2, 8, 'y'' = ', ' ',
    y [2]);
END;


PROCEDURE plot (pars : Paramptr;
                x0, y0, x1, y1 : Citreal;
                clr : Citcolor);
BEGIN
  Load_graph_param (pars);
  Gp_graph_color (clr);
  Moveabs (x0, y0);
  Drawabs (x1, y1)
END;
```

```
BEGIN
  { Set up the windows }
  Define_window (text_w, 15, 2, 65, 5,
    {window position}
    CITYELLOW, CITRED,
    { colours of text   }
    CITWHITE, CITRED,
    { ... of margins }
    CITWHITE, CITBLACK,
    { ... of highlights }
    MARGINS, False);


  Define_window (yt_w, 1, 6, 38, 23,
    {window position}
    CITLIGHTGRAY, CITBLACK,
    {colours of body}
    CITWHITE, CITBLACK,
    { ... of margins }
    PIPYELLOW, AXISRED,
    { ... of axes }
    MARGINS, False);


  Define_window (phase_w, 40, 6, 79, 23,
    {window position}
    CITLIGHTGRAY, CITBLACK,
    {colours of body}
    CITWHITE, CITBLACK,
    { ... of margins }
    PIPYELLOW, AXISRED,
    { ... of axes }
    MARGINS, False);


  { switch on graphics mode ... }
  Graphics_mode;
```

```
{ open windows ... }
Open_window (text_w, 'Report');
Open_window (yt_w, 'y(t) vs t');
Open_window (phase_w, 'Phase plane: y'' vs y');

{set up axes for y(t) vs t and save}
Graph_window (yt_w);
WITH Graphparam DO BEGIN
  xlo := - 0;
  xhi := 30;
  ylo := - 3;
  yhi := 3;
END;
Set_scales;
Draw_axes;
Save_graph_param (yt_pars);

{set up axes for phase plane and save}
Graph_window (phase_w);
WITH Graphparam DO BEGIN
  xlo := - 6;
  xhi := 6;
  ylo := - 6;
  yhi := 6;
END;
Set_scales;
Draw_axes;
Save_graph_param (phase_pars);

initialise;
Display_real (text_w, 2, 1, 6,
  'Van der Pol Oscillator with mu = ', ' ', MU);
print ('Init:');
{Keep track of values from last step}
t_last := t;
y_last := y [1];
yp_last := y [2];
```

```
REPEAT
  oldstep := dt;
  Rkf (ORDER, ABER, RELER, t, dt, DTMIN,
    @f, @y, @yp, nstepsleft, ok);
  (* count and plot results *)
  n := n + 1;
  plot (yt_pars, t_last, y_last, t, y [1],
    CITLIGHTGRAY);
  plot (phase_pars, y_last, yp_last, y [1],
    y [2], CITGREEN);
  {Keep track of values from last step}
  t_last := t;
  y_last := y [1];
  yp_last := y [2];
  Wait (0)
{effect is to test for escape}
UNTIL (nstepsleft = 0) OR NOT ok OR (n = NMAX)
  OR Escape;

{print final values}
print ('Final:');
END.
```

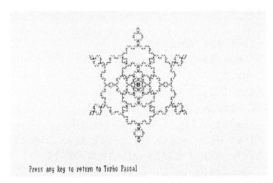

Press any key to return to Turbo Pascal

Figure 7.5 The display produced by program 5-SNOW

⟩ 7.13.5 A fractal curve using basic plot commands

Program 5-SNOW draws a fractal curve which resembles a snowflake. Apart from drawing an attractive and interesting curve, this program

provides another example of the kind of direct control of graphics which
is sometimes needed to produce non-standard results. The program is
also a good example of the use of recursion.

```
{ 5-snow                            RDH 21/9/89  }
{ Level 1 example: basic drawing, fractal curves }

PROGRAM Snowflakes;

  USES
    Cit_core, Cit_prim, Cit_grap;

  CONST
    SIDE = 2.0;

  {length of side of original triangle}
  VAR
    minlength : Citreal;

  PROCEDURE doside (VAR x, y : Citreal;
                        angle, length, hand : Citreal;
                        clr : Citcolor);
  {hand = +1 for right handed, -1 for left}
  BEGIN
    {test for Escape}
    IF Escape THEN
      Exit;
    Wait (0);
    IF Escape THEN
      Exit;
    {test for length at resolution limit}
    IF length < minlength THEN BEGIN
      Gp_graph_color (clr);
      Moveabs (x, y);
      x := x + length * Cos (angle);
      y := y + length * Sin (angle);
      Drawabs (x, y);
      Exit
    END;
```

```
   {if length large enough proceed recursively}
   length := length / 3.0;
   doside (x, y, angle, length, hand, clr);
   angle := angle - hand * Pi / 3.0;
   doside (x, y, angle, length, hand, clr);
   angle := angle + hand * 2 * Pi / 3.0;
   doside (x, y, angle, length, hand, clr);
   angle := angle - hand * Pi / 3.0;
   doside (x, y, angle, length, hand, clr)
END;



{of DoSide}
PROCEDURE triangle (xcentre, ycentre, side,
   theta, hand : Citreal;
                     clr : Citcolor);
   VAR
     r, alpha, x, y : Citreal;
   BEGIN
   r := side / Sqrt (3);
   alpha := theta - Pi / 6.0;
   x := xcentre + r * Cos (alpha);
   y := ycentre + r * Sin (alpha);
   theta := theta + 2 * Pi / 3.0;
   doside (x, y, theta, side, hand, clr);
   theta := theta + 2.0 * Pi / 3.0;
   doside (x, y, theta, side, hand, clr);
   theta := theta + 2.0 * Pi / 3.0;
   doside (x, y, theta, side, hand, clr)
   END;
{of triangle}


{**** Main program ***}
BEGIN


  Graphics_mode;
  Graph_port (12, 1, 68, 25);
```

```
{set up axes}
WITH Graphparam DO BEGIN
  xlo := - 1.5;
  xhi := 1.5;
  ylo := - 1.5;
  yhi := 1.5;
  minlength := 9 * (xhi - xlo) / Screen_wd;
END;

Set_scales;
{Draw_axes; omit except for testing}
triangle (0, 0, SIDE, 0, - 1.0, CITLIGHTCYAN);
triangle (0, 0, SIDE, Pi / 3, - 1.0, CITLIGHTRED)

END.
```

) Chapter 8

) Screen and Text Utilities

The ability to manipulate text is fundamental to the Toolkit. Text itself is a simple object, but the job of placing it on the screen presents some challenges.

One area of difficulty is the variety of screen modes that might be present. The screen can be in graphics or text mode, and there might be 2, 4 or 16 colours available. For the different modes there are different approaches to writing text and assigning foreground and background colours.

Another problem is the limitation of the screen. A display area will normally be no more than 80×25 characters. This is not really sufficient to hold program input and output. Particularly when there are arrays of values, mathematical expressions or text documents, there can easily arise a need to store and manipulate areas of text much greater in extent.

The Toolkit has answers to these problems in the routines contained in the two Level 1 units `Cit_wind` and `Cit_text`. They offer

- an interface between the program and the screen which is independent of mode, and
- text objects which cover a much larger area than the screen.

The interface is a set of routines which write to windows on the screen. As was described in Chapter 6, the communication between user and program can be planned within a windows environment, each window carrying specific information. In principle, for each defined window there is an area of screen and a set of text colours. Text is written to and read from positions within the windows.

Furthermore, the graphics routines described in Chapter 7 will operate within the same window environment. Normally, a graph is drawn within a 'port', a rectangular area of the screen, which can be directly located on the screen by the routine `Graph_port`. Instead, the routine `Graph_window` can be used to link the rectangular area used by the graphics routines to the interior of one of the display windows. Text

291

and graphics can then both be contained within the same display window. Of course, the graphics mode will have to be active at the time, but even then the user writes text to the screen using the same routine calls as might be used for text modes. The Toolkit converts the images automatically.

The Toolkit has text objects which can be used to store numerical arrays and long mathematical expressions. A text object is itself a window, but is not necessarily restricted to the size of the screen. It is a virtual screen of width up to 255 characters and of unlimited height. When text is written to a text object it is placed in heap memory, i.e. the dynamic storage area located above the program in the computer memory. When text is read from a text object, it is copied from this area.

The routines are divided between two units:

- `Cit_wind` contains the low-level windowing routines, which define, save and reload windows, and provide simple operations for writing and reading values within them.
- `Cit_text` contains routines for handling text objects and includes a provision for saving and loading text to and from disc files.

) 8.1   Windows on the screen

Windows have become an important part of the user-interface to a program. Unfortunately, DOS was developed before the use of windows became commonplace. The Toolkit compensates for this by providing a large number of facilities which operate within the DOS environment.

In the commonly used modes there are 80×25 character positions on the screen. At each position there can be a text character, one of 256 available at any time. Each character has a foreground and a background colour.

This is the model that the Toolkit adopts. A copy of the screen is maintained internally within the unit `Cit_wind`. The user will never have to alter this copy directly, but the routines in the unit will modify and read from it, each time making sure the actual video screen corresponds. The advantage of operating this way is that there is protection against accidental corruption. The information stored in memory contains a representation of the screen using an internal coding that is independent of mode, the video screen being refreshed automatically in the best way possible without involving the user with irksome detail.

) *8.1.1 Screen colours*

There are 16 basic colours available for the foreground and background.
They are *black, blue, green, cyan, red, magenta, brown, light grey, dark
grey, light blue, light green, light cyan, light red, light magenta, yellow*
and *white*. In the Turbo Pascal environment the colours are given integer
values 0 to 15, and are available in different combinations in text and
graphics modes.

In the Toolkit there is the type `Citcolor`. Variables of this type
provide a colour mapping valid across different screen modes.

```
CONST N_COLOR_MODES = 4;
TYPE  Citcolor     = ARRAY [1..N_COLOR_MODES] OF
                     byte;
```

Constants are defined of this type in the unit `Cit_core` and describe
mappings onto any of the 2-, 4-, 16- and 256-colour modes. Most work
will use one of the 16 principal colours

CITBLACK	CITBLUE	CITGREEN	CITCYAN
CITRED	CITMAGENTA	CITBROWN	CITLIGHTGRAY
CITDARKGRAY	CITLIGHTBLUE	CITLIGHTGREEN	CITLIGHTCYAN
CITLIGHTRED	CITLIGHTMAGENTA	CITYELLOW	CITWHITE

Colours can be assigned using these names, the Level 1 routines automatically selecting the colour number to suit the screen mode.

Alternatively, colours can be referenced using the integer values normally available in the Turbo Pascal unit `Crt`. There is a global array variable `Citcolorset` available in `Cit_core` which contains all the
colours in their original order. Thus

`Citcolorset[BLACK]` gives the colour `CITBLACK`,

`Citcolorset[BLUE]` gives the colour `CITBLUE`, and so on.

Note that text mode on a PC restricts text background colours; only
the following may be used: *black, blue, green, cyan, red, magenta, brown*
and *light grey*. If you try to use another colour, one of these will be
substituted, according to the following table:

Intended colour	Actual background colour
dark grey	black
light blue	blue
light green	green

Intended colour	Actual background colour
light cyan	cyan
light red	red
light magenta	magenta
yellow	brown
white	light grey

For example, if you try to define a yellow background, you will get a brown one instead. This is a hardware restriction, which the Toolkit is unable to lift.

〉 *8.1.2   Screen representation*

A copy of the text screen is kept in memory in a number of 80 × 25 internal arrays which hold a record of the characters and Toolkit colours at each location. To some extent the arrays are interleaved to mimic the IBM formats for the text screens. This means that text screens can be refreshed quickly. However, all 80-column screen modes, both text and graphic, can be refreshed, the Toolkit colour arrays being used to map the colour values when there are more than or less than 16 colours available.

There are two routines which operate on the whole screen.

The procedure call:

```
Clear_screen;
```

will clear the screen completely, filling the area with space characters.

Clear_screen should only be called before any windows are opened or after all the windows on the screen have been closed. It clears the screen completely; sets the text foreground colour to light grey and the text background colour to black; and places the cursor at the top left of the screen. The (text or graphics) screen mode remains unchanged.

〉 *Procedure Clear_screen*

Routine name:	**Clear_screen**
Kept in unit:	**Cit_wind**
Purpose:	Clear the screen

The unit name **Cit_wind** must appear in the uses clause at the head of your program.

Signature:

```
procedure Clear_screen;
```

The procedure call

```
Refresh_text_screen;
```

will refresh the screen after a mode change, or some other process (say, an exit to a DOS shell) has corrupted the screen, and you wish to redraw it.

> *) Procedure Refresh_text_screen*

Routine name: **Refresh_text_screen**
Kept in unit: **Cit_wind**
Purpose: Redraw complete text screen from backup copy

The unit name **Cit_wind** must appear in the uses clause at the head of your program.

Signature:

```
procedure Refresh_text_screen;
```

> *) 8.1.3 Defining and locating windows*

The Toolkit windows themselves are defined by variables of the type **Citwindow**. Because this type is fundamental, it is defined in the unit **Cit_core**, rather than in **Cit_wind**, the windows unit.

A call to the routine **Define_window** is the fundamental way of specifying where a window is placed, what its colours are, and how its images are to be displayed and stored. As so many attributes have to be set there have to be several parameters; but these are quite easy to understand.

Consider a typical call of **Define_window**:

```
Define_window (window_variable,  { of type Citwindow }
    x1, y1, x2, y2,              { coordinates }
    textfore, textback,          { text colours }
    margfore, margback,          { margin colours }
    highfore, highback,          { highlight colours }
    window_type,                 { margins present? }
    buffer_image);               { save screen? }
```

There are 13 parameters in all; let us consider them in order:

- **window_variable** contains all the information about the window;
  during the call, **window_variable** will be assigned all the information about margins, colours, and positions. (In fact, the routine **Define_window** merely initialises the various components of
  window_variable.)
- **x1, y1, x2** and **y2** give the position and size of the window. The screen
  is considered to be a grid of points numbered 1 to 80 in the horizontal
  (x) direction, and 1 to 25 in the vertical (y) direction*. Note that the
  points are counted to the right horizontally (just as with the x-axis of
  a graph), but that the points are counted *downwards* vertically (the
  opposite direction from the y-axis of a normal graph). The window
  is defined by giving the coordinates of any pair of diagonally opposite points. For example, if the window has corners at coordinates
  $(1,1)$, $(20,1)$, $(20,10)$ and $(1,10)$, then either of the pairs of corners
  $(1,1); (20,10)$ or $(1,10); (20,1)$ would define the window. We refer to
  these pairs of points as the *defining corners* of the window; these are
  shown in figure 8.1 below.

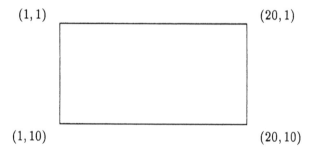

Figure 8.1   Defining corners of a window

- **textfore** and **textback** are the foreground and background colours,
  respectively, of text in the window. The unit **Cit_core** defines all the
  colours available: **CITLIGHTGRAY**, for instance, represents the colour
  light grey.
- **margfore** and **margback** define the foreground and background
  colours, respectively, of the margins.

---

* Just as if you were dealing with a text screen.

- **highfore** and **highback** define the foreground and background colours, respectively, of 'highlighted' information in the window; this is used, in the main, by the menu routines in **Cit_menu**.

- **window_type** must be one of the scalar constants **MARGINS** or **NOMARGINS**. This controls, in the obvious way, whether or not the window has margins. Note that the size of the window, as given by **x1**, **y1**, **x2** and **y2**, *includes* the margins. Margins, if present, take up two columns (one on either side of the window) and two rows (one at the top, and one at the bottom). Imagine that the window's lower left-hand corner is at (x_l, y_l), and its upper right-hand corner at (x_h, y_h). Then, if **window_type** is equal to **MARGINS**, the lower left-hand corner of the area of the window available to the user will be at $(x_l + 1, y_l - 1)$, and the upper right-hand corner at $(x_h - 1, y_h + 1)$. **Define_window** behaves in this way, so that the area the window takes up on the screen is always the same, regardless of whether or not the window has margins. If there is not enough room for the window to contain margins (it has fewer than two rows, or fewer than two columns), **Define_window** behaves as though **window_type** had been **NOMARGINS**.

- **buffer_image** is a Boolean, which controls whether or not the window 'remembers' the screen which it overwrites. If **buffer_image** is **false**, then when the window is opened on the screen the Toolkit will not save what was there originally; this is the correct behaviour for the windows at the 'base' of a cleared screen, of course. If **buffer_image** is **true**, the Toolkit will save the portion of the screen over which the window lies; when the window is closed, the screen will be restored to its former state. This behaviour is correct for things like pop-up menus or help windows.

You should note that setting **buffer_image** to **true** will increase the amount of memory a program using the Toolkit requires; but it is a very useful facility and there are safeguards built in to save images to disc if memory space becomes scarce. You can minimise the amount of additional memory required, yet still keep fast window changes by making the pop-up window as small as possible.

Define_window associates various graphics colours with the window you define. Axes are drawn in red; markers on axes ('pips') in yellow; the labels in light grey; and lines in white. These default colours may be changed by calling **Set_graph_colours**.

Let us try to enliven this rather dry description with a few example

windows. We'll consider the set of three windows you might wish to use
for a simple program: a pair of windows which together occupy the whole
screen, and a third, which acts as a window in which help information
appears. (We might envisage that one of the pair of windows is used for
parameters to a calculation, and the other for its results.) We shall leave
the bottom line (line 25) blank, so that a status window can appear in
it, if we wish.

Imagine that we divide the screen vertically, and wish the results window to occupy three-quarters of the screen, and the parameters window
the remainder. The larger window will have its upper left-hand corner
at $(1, 1)$, and its lower right-hand corner at $(60, 24)$.

Let us assume that we don't want the window to have margins, and
all the information it contains is displayed as yellow on brown (that is,
the foreground colour is yellow, and the background colour brown). As
this window is always going to be present on the screen, there is no need
to remember what the screen underneath it looked like, so we can set
the parameter **buffer_image** to **false**. We are now in a position to
define the first window by the following Pascal program:

```
program define_windows;
  uses Cit_core, Cit_wind;
var w1, w2, hw : Citwindow;
begin
  Define_window (w1,    { window }
  1, 1, 60, 24,         { coordinates }
  CITYELLOW, CITBROWN, { text colours }
  CITYELLOW, CITBROWN, { margin colours }
  CITYELLOW, CITBROWN, { highlight colours }
  NOMARGINS,            { no margins }
  false);               { forget screen under window }
end.
```

All the comments in the program above are, of course, strictly unnecessary; but are there to remind us of the meaning of the parameters.

Let us now define the second window, which lies to the right of the first
and takes up the remainder of the screen. This makes its two defining
corners $(61, 1)$ and $(80, 24)$. We shall give this window margins; whilst
this does not change the call of **Define_window**, it does mean that the
area available for us to use is defined by the corners $(62, 2)$ and $(79, 23)$;
the margins take up two rows, and two columns, at the edges of the
window.

Let us assume that we want this window to have a cyan background, with white text; then we can define it by inserting the following into the **define_windows** program above:

```
Define_window (w2,    { window }
   61, 1, 80, 24,     { coordinates }
   CITWHITE, CITCYAN, { text colours }
   CITWHITE, CITCYAN, { margin colours }
   CITWHITE, CITCYAN, { highlight colours }
   MARGINS,           { with margins }
   false);            { forget screen under window }
```

Finally, we have to define the window for 'help' information. As we presumably want this to 'pop-up' when needed, and disappear otherwise, we must arrange that it preserves the screen over which it lies, and set **buffer_image** to **true**. Let us put the window in the centre of the screen, occupying 11 lines; and let us give it a width of 40 columns. These requirements give the defining corners of the window as $(21, 8)$ and $(60, 18)$. Let us give the window margins, and have the text given in black on a green background. We can define **hw** to do all this by inserting the following Pascal into the **define_windows** program above:

```
Define_window (hw,    { window }
   21, 8, 60, 18,     { coordinates }
   CITBLACK, CITGREEN, { text colours }
   CITBLACK, CITGREEN, { margin colours }
   CITBLACK, CITGREEN, { highlight colours }
   MARGINS,           { with margins }
   true);             { save screen under window }
```

These windows are all defined, and displayed, in the working example to accompany **Define_window**.

⟩ 8.1.4   *Working example of Define_window*

Filename: **CHAP8\8-1DEFIN.PAS**
To run the working example see *Guide to running the software*.

We can now summarise the call.

⟩ *Procedure Define_window*

Routine name: `Define_window`
Kept in unit: `Cit_wind`
Purpose: Define window coordinates, colours, margins, and imaging

The unit name `Cit_wind` must appear in the uses clause at the head of your program.

Signature:

```
procedure Define_window (
  VAR w                   : Citwindow;
  x1, y1, x2, y2          : integer;
  textcolor, textbkcolor,
  margcolor, margbkcolor,
  highcolor, highbkcolor  : Citcolor;
  wndtype                 : Windowtype;
  bufferedimage           : boolean);
```

Input parameters for `Define_window`:

`x1, y1, x2, y2`:   Position of window
`textcolor, textbkcolor`: Text colours
`margcolor, margbkcolor`: Margin colours
`highcolor, highbkcolor`: Highlight colours
`wndtype`:   Window type
`bufferedimage`:   Image buffer flag

Output parameters for `Define_window`:

`w` is set to the new window, with the appropriate settings.

There is a null window `No_window` of type `Citwindow` declared in the unit `Cit_wind`. It is useful if a window is to be completely cancelled. The statement

```
w := No_window
```

will set the window variable `w` to be a null window. Alternatively, the window type (which, for window variable `w`, is contained in the window variable `w.Wtype`) can be switched to `NULL` and the window suspended temporarily. It would be restored by setting the type variable to the value `MARGINS` or `NOMARGINS`, as set when the window was initialised.

The position of a window may be changed at any time by calling `Locate_window`. This will not affect anything which appears on the screen until the window is opened or painted.

⟩ *8.1.5* *Working example of Locate_window*

Filename: `CHAP8\8-1LOCAT.PAS`
To run the working example see *Guide to running the software.*

⟩ *Procedure Locate_window*

Routine name: `Locate_window`
Kept in unit: `Cit_wind`
Purpose: Change the coordinates of a window

The unit name `Cit_wind` must appear in the uses clause at the head of your program.

Signature:

```
procedure Locate_window (var w  : Citwindow;
                         x1, y1,
                         x2, y2 : integer);
```

Input parameters for `Locate_window`:

Call:	`Locate_window (w, x1, y1, x2, y2)`
`w`:	Window variable
`x1, y1`:	Coordinates of one defining point
`x2, y2`:	Coordinates of the other defining point

Output parameters for `Locate_window`:

`w` is modified to contain the new coordinates.

For example, if `w` is a window variable, and we wish to reposition it so that its defining corners are at $(10, 20)$ and $(30, 40)$, the correct call of `Locate_window` is:

```
Locate_window (w, 10, 20, 30, 40);
```

If the window has margins, but the window specified by `Locate_window` is too small to contain them (it has fewer than two columns, or fewer than two rows), then `Locate_window` converts the window into one without margins.

The routine `Color_window` allows you to change all the window colours (with the exception of the graphics colours, for which you use `Set_graph_colours`) at the same time. It modifies the window so that the new colours are used immediately.

) *Procedure Color_window*

Routine name: `Color_window`
Kept in unit: `Cit_wind`
Purpose: Change the coordinates of a window

The unit name `Cit_wind` must appear in the uses clause at the head of your program.

Signature:

```
procedure Color_window (VAR w        : Citwindow;
                        textcolor,
                        textbkcolor,
                        margcolor,
                        margbkcolor,
                        highcolor,
                        highbkcolor : Citcolor);
```

Input parameters for `Color_window`:

Call:	`Color_window (w, tf, tb, mf, mb, hf, hb)`
w:	Initialised window variable
tf, **tb**:	Text foreground and background colours, respectively
mf, **mb**:	Margin foreground and background colours, respectively
hf, **hb**:	Highlight foreground and background colours, respectively

Output parameters for `Color_window`:

w is modified to contain the new colours.

Imagine that we wish to change the colours of the help window introduced above in the example of `Define_window`, interchanging the foreground and background colours (so that information appears green on black, rather than black on green). This is done by the following call:

```
Color_window (w,
    CITGREEN, CITBLACK,   { text colours }
    CITGREEN, CITBLACK,   { margin colours }
    CITGREEN, CITBLACK);  { highlight colours }
```

Color_window changes all the colours in a window at once, which is useful when you want to make major changes. However, it is rather tedious to use if you merely wish, say, to change the colours in the margin without affecting any others. This is where the following three routines are useful. Set_text_colors resets the text foreground and background colours, and leaves the others unchanged. Set_margin_colors changes the margin colours only; and Set_high_colors the highlight colours only. The parameters are similar for the three routines.

⟩ *Procedure Set_text_colors*

Routine name: Set_text_colors
Kept in unit: Cit_wind
Purpose: Change window's text colours

The unit name Cit_wind must appear in the uses clause at the head of your program.

Signature:

```
procedure Set_text_colors (var w  : Citwindow;
                           fore,
                           back   : Citcolor);
```

Input parameters for Set_text_colors:

Call: Set_text_colors (w, tf, tb);
w: Initialised window variable
tf, tb: Text foreground and background colours, respectively

Output parameters for Set_text_colors:

w is modified to contain the new colours.

) *Procedure Set_ margin_colors*

Routine name: `Set_margin_colors`
Kept in unit: `Cit_wind`
Purpose: Change window's margin colours

The unit name `Cit_wind` must appear in the uses clause at the head of
your program.

Signature:

```
procedure Set_margin_colors (var w  : Citwindow;
                             fore,
                             back   : Citcolor);
```

Input parameters for `Set_margin_colors`:

Call: `Set_margin_colors (w, mf, mb);`
mf, mb: Margin foreground and background colours,
                  respectively

) *Procedure Set_ high_colors*

Routine name: `Set_high_colors`
Kept in unit: `Cit_wind`
Purpose: Change window's highlight colours

The unit name `Cit_wind` must appear in the uses clause at the head of
your program.

Signature:

```
procedure Set_high_colors (var w  : Citwindow;
                           fore,
                           back   : Citcolor);
```

Input parameters for `Set_high_colors`:

Call: `Set_high_colors (w, hf, hb);`
hf, hb: Highlight foreground and background colours,
                  respectively

For example, if `w` is a window variable, we can change the highlight
foreground and background colours to yellow and red, respectively, by
the Pascal statement:

```
Set_high_colors (w, CITYELLOW, CITRED);
```

The routine **Define_window** sets the colours for graphs by default, rather than explicitly. Axes are drawn in red, with yellow markers ('pips'), and light-grey labels; and lines on graphs are drawn in white. The routine **Set_graph_colours** enables you to change these colours as you wish.

⟩ *Procedure Set_graph_colors*

Routine name: **Set_graph_colors**
Kept in unit: **Cit_wind**
Purpose: Change window's graph colours

The unit name **Cit_wind** must appear in the uses clause at the head of your program.

Signature:

```
procedure Set_graph_colors (var w  : Citwindow;
                            pipcolor,
                            axiscolor,
                            labelcolor,
                            linecolor : Citcolor);
```

Input parameters for Set_graph_colors:

Call: **Set_graph_colors (w, pipc, axisc, labc, linec);**
w: Window variable
pipc: Defines the 'pip' colour
axisc: Defines the axis colour
labc: Defines the label colour
linec: Defines the line colour

The call modifies the colours stored in the window **w**: the 'pip' colour is set to **pipc**, the axis colour to **axisc**, the label colour to **labc**, and the line colour to **linec**.

Output parameters for Set_graph_colors:

w is modified to contain the new colours.
     For example, the statement:

```
Set_graph_colors (w, CITBLUE, CITYELLOW,
    CITRED, CITGREEN);
```

would cause graphs plotted in the window stored in **w** to have blue pips on yellow axes, with red labels and green lines.

⟩ *8.1.6   Opening and closing a window*

Windows which have been defined by **Define_window** are not displayed until they are *opened*.

If the window has a **bufferedimage** set **true**, it preserves the portion of the screen which it will cover. If there is room, the full window image is transferred to heap memory; otherwise, an attempt is made to store it in a temporary disc file. The files so generated are given the extension **.WND**, and are deleted as soon as the window is closed and the image restored. To make sure that the whole of the memory is not filled with window images, to the exclusion of all other memory uses, a threshold is set beyond which images are sent to disc file. The threshold is determined by the value of the global variable **Non_window_reserve** (of type **word**) which is in the unit **Cit_wind**. It is set initially to the value 4096, but can be changed by the user.

If there is already a buffered image in memory corresponding to the window, then this image is discarded before attempting to save the new one.

If it is not possible to save the image in memory or on file, then the window is rendered 'inactive' and **Errorflag** set **true**. Text cannot be written to an inactive window. Thus, in this error state the screen will not be corrupted, the user being given an opportunity to recover. The only way a window can be made 'active' (except, of course, by deliberately changing the window flag **Wactive**, a practice which is not recommended) is by opening the window with adequate memory for any buffered image.

If the window has margins, these are drawn when a window is opened, and an optional title written in the top left-hand corner. Then, the window is cleared to its background text colour.

⟩ *8.1.7   Working example of Open_window*

Filename: **CHAP8\8-1LOCAT.PAS**
To run the working example see *Guide to running the software.*

⟩ *Procedure Open_window*

Routine name:	**Open_window**
Kept in unit:	**Cit_wind**
Purpose:	Open a window which has been initialised

The unit name `Cit_wind` must appear in the uses clause at the head of your program.

Signature:

```
procedure Open_window (var w : Citwindow;
                       lline : Linestring);
```

Input parameters for Open_window:

Call:	`Open_window (w, title);`
`w`:	Window variable
`title`:	Title string for the window

Output parameters for Open_window:

If `buffer_image` was true when `Define_window` was called, `w` is modified to hold a copy of the portion of the screen overwritten by the window.

For example, if `hw` is the help window we defined above using `Define_window`, the Pascal statement:

```
Open_window (hw, 'Help window');
```

'opens' the help window, and gives it the title `Help window`. You will note (from the call of `Define_window` which created `hw`) that the window is defined to save any portion of the screen over which it lies, to have margins, and have a black foreground on a green background. The call above, therefore, saves the part of the screen which will be 'covered' by the window; clears the window area to green; and surrounds the window with a black line as a margin (the foreground margin colour being black). Finally, the title "Help window" is displayed in the top left-hand corner of the margin.

When work in a window has finished, the window is closed. This is achieved with a call to the routine `Close_window`, which complements the routine `Open_window`. If the window was opened with `bufferedimage` set to `true` (that is, if it preserved the portion of screen it overwrote), then the Toolkit automatically restores the portion of screen affected to its original state, and the image is removed from the heap (a buffered image may be restored only once). Otherwise, `Close_window` does nothing.

) *Procedure Close_window*

Routine name: `Close_window`

Kept in unit: `Cit_wind`
Purpose: Close a window opened by `Open_window`

The unit name `Cit_wind` must appear in the uses clause at the head of your program.

Signature:

```
procedure Close_window (var w : Citwindow);
```

Input parameters for `Close_window`:

Call: `Close_window (w);`
`w`:   Window variable

Output parameters for `Close_window`:

For example, when we have finished displaying help information in the window defined by `hw`, the call:

```
Close_window (hw);
```

will cause the portion of screen overwritten by the help window to be restored to its former state. This is exactly what is wanted from 'pop-up' windows, of course.

) 8.1.8   Window images in memory

The Toolkit provides facilities for saving and restoring window images. The routines **Save_window** and **Load_window** quickly copy the graphics and text images to and from heap memory. The routine **Free_window** clears an image from memory.

Windows are saved to, and may be reloaded from, variables of type **Wimageptr**, which is defined in unit **Cit_core**. If `w` is a window variable, and `wcopy` a variable of type **Wimageptr**, then the Pascal statement:

```
Save_window (w, wcopy);
```

saves the contents of the window `w` in `wcopy`. To restore the window to this state, you merely need:

```
Load_window (w, wcopy);
```

which restores the window **w** to the state recorded in **wcopy**. You can do this whenever you wish, and as often as you wish, provided that the window has not been closed.

If the size of the window, or the screen mode, has changed between saving and loading then the Toolkit automatically adjusts as best it can.

⟩ *Procedure Save_window*

Routine name: **Save_window**
Kept in unit: **Cit_wind**
Purpose: Save a copy of a window

The unit name **Cit_wind** must appear in the uses clause at the head of your program.

Signature:

```
procedure Save_window (w     : Citwindow;
                       var b : Wimageptr);
```

Input parameters for Save_window:
Call: **Save_window (w, b);**
w: Window variable
b: Variable of type **Wimageptr**

Output parameters for Save_window:
b is modified to contain a copy of the current contents of the window.

Error indicators for Save_window:
An error will occur if there is insufficient memory to save the window. Then, **Errorflag** is set **true** and the variable b is set to **NIL**. When this happens you can decide either to stop, or proceed and save the image on disc. (When **Save_window** is called by **Open_window**, the decision is taken to proceed.)

⟩ *Procedure Load_window*

Routine name: **Load_window**
Kept in unit: **Cit_wind**
Purpose: Reload a window from a saved copy

The unit name **Cit_wind** must appear in the uses clause at the head of your program.

Signature:

```
procedure Load_window (w : Citwindow;
                       b : Wimageptr);
```

Input parameters for Load_window:

Call: Load_window (w, b);
w: Window variable
b: Value of type **Wimageptr**, created by **Save_window**

Windows can require a large amount of memory, especially if you are using a VGA. Once you have finished with a saved window image, it is wise to dispose of it (and free the memory used) by calling **Free_window**. This routine takes only one parameter: the window image, stored in a variable of type **Wimageptr**.

If **wcopy** contains a window image, then:

```
Free_window (wcopy);
```

will release the memory occupied by the window image.

) Procedure Free_window

Routine name: **Free_window**
Kept in unit: **Cit_wind**
Purpose: Free the store occupied by a window image

The unit name **Cit_wind** must appear in the uses clause at the head of your program.

Signature:

```
procedure Free_window (var b : Wimageptr);
```

Input parameters for Free_window:

Call: Free_window (b);
b: Variable of type **Wimageptr**

Output parameters for Free_window:

b is set to **NIL**.

⟩ *8.1.9   Window images on disc files*

The Toolkit also provides facilities for saving and restoring window images to and from disc files. The routines **Write_window** and **Read_window** copy the graphics and text images to and from disc.

If **w** is a window variable, and **fname** a legal file name then the Pascal statement:

```
Write_window (w, fname);
```

saves the contents of the window **w** to file **fname**. To restore the window, you merely need:

```
Read_window (w, fname);
```

You can do this whenever you wish, and as often as you wish, provided that the window has not been closed or changed.

The files are given an extension **.WND** by default.

If the screen mode has changed between writing and reading, the Toolkit automatically adjusts as best as it can. However, the Toolkit is not able to cope if the window size has changed.

⟩ *Procedure Write_window*

Routine name: **Write_window**
Kept in unit: **Cit_wind**
Purpose: Write a copy of a window to file

The unit name **Cit_wind** must appear in the uses clause at the head of your program.

Signature:

```
procedure Write_window (w      : Citwindow;
                        fname : Linestring);
```

Input parameters for Write_window:

Call: **Write_window (w, b);**
w: Window variable
fname: Legal file name

Error indicators for Write_window:
Errorflag is set **true** if there is a file error.

⟩ *Procedure Read_window*

Routine name:	**Read_window**
Kept in unit:	**Cit_wind**
Purpose:	Read a window image from file

The unit name **Cit_wind** must appear in the uses clause at the head of your program.

Signature:

```
procedure Read_window (w     : Citwindow;
                       fname : Linestring);
```

Input parameters for Read_window:

Call:	**Read_window (w, fname);**
w:	Window variable
fname:	Name of an existing file

⟩ *8.1.10 Clearing a window and resetting margins*

The routine **Clear_window** clears a window by colouring its text area with the text background colour; the margins are untouched. The current text position within the window is set to the top-left corner. The complementary routine **Margin_window** redraws the margins using current margin colours and leaves the text area untouched.

If **w** is a window variable, then the call:

```
Clear_window (w);
```

will clear the window defined by the variable **w** to its text background colour.

⟩ *Procedure Clear_window*

Routine name:	**Clear_window**
Kept in unit:	**Cit_wind**
Purpose:	Clear a window

The unit name **Cit_wind** must appear in the uses clause at the head of your program.

Signature:

```
        procedure Clear_window (var w : Citwindow);
```

Input parameters for Clear_window:

Call: Clear_window (w);

w: Window variable

⟩ *Procedure Margin_window*

Routine name: Margin_window
Kept in unit: Cit_wind
Purpose: Redraw the margin

The unit name Cit_wind must appear in the uses clause at the head of your program.

Signature:

```
        procedure Margin_window (w      : Citwindow;
                                 title : Linestring);
```

Input parameters for Margin_window:

Call: Margin_window (w, title);

w: Window variable

title: Legal file name

This routine is useful if you wish to change the window title, or if the margin colours are changed when a window is activated.

⟩ *8.1.11 Painting a window*

Paint_window completely redraws a text window using colours supplied by the user.

One use of this is to provide visual evidence that a window is deactivated: for example, a menu window might be wholly repainted in duller colours once the user has finished choosing from it.

⟩ *Procedure Paint_window*

Routine name: Paint_window
Kept in unit: Cit_wind
Purpose: Redraw text window in new colours

The unit name Cit_wind must appear in the uses clause at the head of your program.

Signature:

```
procedure Paint_window (w              : Citwindow;
                        txtcolor,
                        txtbkcolor,
                        margcolor,
                        margbkcolor : Citcolor);
```

Input parameters for Paint_window:

Call:	Paint_window (w, tf, tb, mf, mb);
w:	Window to be painted
tf, tb:	Text foreground and background colours, respectively
mf, mb:	Margin foreground and background colours, respectively

) 8.1.12   Scrolling windows

Scroll_window_n scrolls the text in a given window by a number of lines.

) Procedure Scroll_window_n

Routine name:	Scroll_window_n
Kept in unit:	Cit_wind
Purpose:	Scroll a window

The unit name Cit_wind must appear in the uses clause at the head of your program.

Signature:

```
procedure Scroll_window_n (var w  : Citwindow;
                           nlines : integer);
```

Input parameters for Scroll_window_n:

Call:	Scroll_window_n (w, nlines);
w:	Window to be painted
nlines:	Number of lines by which to scroll (may be negative)

For example, if **w** is a window, then:

        Scroll_window_n (w, n);

scrolls the text in the window by **n** lines. **n** may be positive (in which case the text moves up), or negative (in which case the text moves down). If |**n**| is larger than the number of lines in the window, the window is cleared, and the cursor moved to the beginning of the bottom line of the window (if **n** is positive) or the beginning of the top line of the window (if **n** is negative).

⟩ 8.1.13   Routine Scroll_window

Scroll_window scrolls a window up by one line:

        Scroll_window (w);

is identical to:

        Scroll_window_n (w, 1);

⟩ Procedure Scroll_window

Routine name: **Scroll_window**
Kept in unit: **Cit_wind**
Purpose: Scroll a window by one line

The unit name **Cit_wind** must appear in the uses clause at the head of your program.

Signature:

        procedure Scroll_window (var w   : Citwindow);

Input parameters for Scroll_window:
Call: Scroll_window (w);
w: Window to be scrolled

⟩ 8.1.14   Examining a window character

The function **Text_char_at** returns the character at a given row and column within a window. If **w** is a window, then:

        Text_char_at (w, 2, 3);

returns the character at column 2, row 3 of the window.

) *Function Text_char_at*

Routine name:	**Text_char_at**
Kept in unit:	**Cit_wind**
Purpose:	Return character in text window

The unit name **Cit_wind** must appear in the uses clause at the head of your program.

Signature:

```
function Text_char_at (w       : Citwindow;
                       col, row : integer) : char;
```

Input parameters for Text_char_at:

Call:	c := Text_char_at (w, col, row);
w:	Window to be read
col:	Column within the window
row:	Row within the window

Output parameters for Text_char_at:

Text_char_at returns the character at the given column and row within the window.

Error indicators for Text_char_at:

If **col** and/or **row** lie outside the window, **Text_char_at** returns a character whose ordinal value is zero (which can be written in Turbo Pascal as **#0**).

) *8.1.15   Writing a string to a window*

The procedure **Write_string** writes a string at a given point within a window. Its parameters are the window in which the string is to be displayed, the column and row at which it is to start, and the string itself. The string is displayed in the window's text foreground colour, against a background of the window's text background colour.

) *8.1.16   Working example of Write_string*

Filename: **CHAP8\8-1LOCAT.PAS**

To run the working example see *Guide to running the software.*

⟩ *Procedure Write_string*

Routine name: `Write_string`
Kept in unit: `Cit_wind`
Purpose: Display a string within a window, using text colours

The unit name `Cit_wind` must appear in the uses clause at the head of your program.

Signature:

```
procedure Write_string (var w    : Citwindow;
                        col, row : integer;
                        lline    : Linestring);
```

Input parameters for `Write_string`:

Call: `Write_string (w, col, row, lline);`
`w`:   Window to be read
`col`:   Column within the window
`row`:   Row within the window
`lline`:   String to be written

For example,

```
Write_string (w, 2, 3, 'hello, there!');
```

would write the message `hello, there!` at column 2, row 3 of window `w`. If the string is too long to fit in the window (for example, if the window in the above example had only five columns), it is shortened to fit.

`Write_string` has the additional property that, if the row given, r, is negative or zero, it *adds* $|r|$ to the current position in the row, and writes the string there. Similarly, if the column, c, given is negative, it adds $|c|$ to the current column, and writes the string there. Thus the two Pascal statements:

```
Write_string (w, 2, 3, 'hello there');
Write_string (w, 0, 0, ', my friend');
```

would write:

```
hello there, my friend
```

in the window `w` (assuming that there is space to contain the message).

The procedure **Highlight_string** is identical to **Write_string**, with the single exception that it writes the string in the highlight foreground colour, with a background of the highlight background colour.

This is useful for displaying important information (such as error messages) in the window.

⟩ *Procedure Highlight_string*

Routine name: **Highlight_string**
Kept in unit: **Cit_wind**
Purpose: Display a string within a window, using highlight
                  colours

The unit name **Cit_wind** must appear in the uses clause at the head of your program.

Signature:

```
procedure Highlight_string (var w    : Citwindow;
                            col, row : integer;
                            lline    : Linestring);
```

Input parameters for Highlight_string:
Call: Highlight_string (w, col, row, lline);
                  Parameters as for **Write_string**.

⟩ *8.1.17   Painting a string*

Paint_string paints only a *line* on the screen within a window. It changes the colours on the screen and in the screen copy. The facility is used when highlighting the current menu option, or when flashing a line of text.

⟩ *Procedure Paint_string*

Routine name: **Paint_string**
Kept in unit: **Cit_wind**
Purpose: Paint a string within a window

The unit name **Cit_wind** must appear in the uses clause at the head of your program.

Signature:

```
procedure Paint_string (w          : Citwindow;
                        col, row   : integer;
                        len        : integer;
                        txtcolor,
                        txtbkcolor : Citcolor);
```

Input parameters for Paint_string:

Call:	`Paint_string (w, col, row, len, tf, tb);`
w:	Window to be read
col:	Column within the window
row:	Row within the window
len:	Length of string to be redrawn
tf, tb:	Text foreground and background colours, respectively

For example,

```
Paint_string (w, 2, 3, 4, CITBLACK, CITGREEN);
```

would repaint the four characters in w, starting at column 2, row 3, to have a foreground colour of black, and a background colour of green. If the string does not lie completely within the window (for example, the start row or column lies outside the window, or the string is too long to fit into the window), `Paint_string` does nothing.

⟩ 8.1.18 *Reading strings and values from within a window*

The procedure `Read_string` reads a string from the keyboard, displaying the result in a portion of a window. The string may have a default value, and may be edited using the usual PC keys. Similarly, the routines `Read_integer` and `Read_real` read integer and real values from the keyboard.

⟩ *Procedure Read_string*

Routine name:	`Read_string`
Kept in unit:	`Cit_wind`
Purpose:	Read a string from the keyboard, displaying it within a window

The unit name `Cit_wind` must appear in the uses clause at the head of your program.

Signature:

```
procedure Read_string (var w     : Citwindow;
                        col, row : integer;
                        default  : Linestring;
                        var s    : Linestring);
```

Input parameters for Read_string:

Call:	Read_string (w, col, row, default, s);
w:	Window to be read
col:	Column within the window
row:	Row within the window
default:	Default value for string

Output parameters for Read_string:

s:	String contains the result

For example, if iw is the window to be used for string input, the Pascal statement:

```
Read_string (iw, 10, 1, 'normal string', s);
```

would input a string, starting in column 10, row 1 of the window specified by iw.

When **Read_string** is called, the user can immediately press Enter, in which case **Read_string** behaves as though the user had typed the default string. Otherwise, the user can type in another value for the string, or use the normal PC editing keys. Before the user presses any keys, the cursor is positioned at the start of the default string; thereafter:

- The right-arrow key → moves the cursor one character to the right.
- The left-arrow key ← moves the cursor one character to the left.
- Home moves the cursor to the start of the string.
- End moves the cursor to the place just to the right of the end of the string.
- Del deletes the character under the cursor, if any.
- Ins switches between insert and overtype mode; immediately after **Read_string** has been called, overtype mode is in effect.
- The rubout key (←, immediately above Enter) deletes the character to the left of the cursor, and moves the cursor one place left.
- Enter finishes the input process: the string displayed in the window is assigned to the parameter s, and **Read_string** returns.
- Esc or Break have the same effect as pressing Enter.

Read_integer is almost identical to **Read_string**, except that it inputs an integer, rather than a string.

) *8.1.19   Working example of Read_integer*

Filename: **CHAP8\8-1LOCAT.PAS**
To run the working example see *Guide to running the software.*

) *Procedure Read_ integer*

Routine name: **Read_ integer**
Kept in unit: **Cit_wind**
Purpose: Read an integer from the keyboard, displaying it
                  within a window

The unit name **Cit_wind** must appear in the uses clause at the head of your program.

Signature:

```
procedure Read_integer (var w    : Citwindow;
                        col, row : integer;
                        default  : Linestring
                        var n    : integer);
```

Input parameters for Read_integer:

Call:	**Read_integer (w, col, row, default, n);**
w:	Window to be read
col:	Column within the window
row:	Row within the window
default:	String containing default value for integer

Output parameters for Read_integer:

n:	Integer to contain result

Error indicators for Read_integer:

The string entered is converted to an integer value using the Pascal procedure **Val**. If an error code is returned, then the flag **Errorflag** is set **true**.

The one thing to beware of is that the default value for the integer is provided by a *string*, not an integer.

The routine **Read_real** is also based on the routine **Read_string** and behaves similarly to the routine **Read_integer**.

⟩ *Procedure Read_ real*

Routine name:	**Read_real**
Kept in unit:	**Cit_wind**
Purpose:	Read a real from the keyboard, displaying it within a window

The unit name **Cit_wind** must appear in the uses clause at the head of your program.

Signature:

```
procedure Read_real (var w    : Citwindow;
                     col, row : integer;
                     default  : Linestring
                     var r    : Citreal);
```

Input parameters for Read_real:

Call:	**Read_real (w, col, row, default, s);**
w:	Window to be read
col:	Column within the window
row:	Row within the window
default:	Default value for real

Output parameters for Read_real:

r:	Real to contain result

Error indicators for Read_real:

The string entered is converted to a **Citreal** value using the Pascal procedure **Val**. If an error code is returned, then the flag **Errorflag** is set **true**.

Again, beware of the fact that the default value for the real is provided by a *string*, not a **Citreal** value.

⟩ **8.2   Manipulating text objects**

A general text object has been introduced into the Toolkit to store arrays and expressions which might occupy several lines and which require a text area greater than the 80 × 25 area on the screen.

Text objects are held in variables of type `Textptr`, which is defined in the unit `Cit_core`. They consist of an arbitrary number of lines of text; each line is represented by a Turbo Pascal string, and thus is restricted to 255 characters, but there is no limit on the number of lines present. A text object may therefore hold a string of infinite length*.

The type `Textptr` is a Pascal pointer type; text objects may therefore be assigned the Pascal pointer value `NIL`. This corresponds to a text object containing no lines; we shall refer to this as the *null* text object.

Because text objects are used throughout the Toolkit, the fundamental definitions and some basic routines are kept in the unit `Cit_core`, rather than in the unit `Cit_text` with the main set of the text routines.

⟩ 8.2.1   Creating and removing text objects

A text object is created by the function `New_text`, which converts a string into a text object, and returns the result. More accurately, `New_text` creates a text object consisting of only one line, which contains the string it was given.

⟩ 8.2.2   Working example of New_text

Filename: `CHAP8\8-2NEW.PAS`
To run the working example see *Guide to running the software.*

⟩ *Function New_text*

Routine name: `New_text`
Kept in unit: `Cit_core`
Purpose: Create a new one-line text object

The unit name `Cit_core` must appear in the uses clause at the head of your program.

Signature:

```
function New_text (s : string) : Textptr;
```

Input parameters for New_text:

Call: `New_text (str)`
str: String giving the one line of text object

---

* In practice, of course, the amount of memory available on the computer will limit the size of the text object which can be stored.

Output parameters for New_text:
The function returns the new text object, consisting of one line, equal to **str**.

Error indicators for New_text:
If there is insufficient memory, **Errorflag** is set **true** and the function returns the **NIL** pointer.

As an example of **New_text**:

```
program show_new_text;
  uses Cit_core;
var
  t : Textptr;
begin
  t := New_text ('hello, world!')
end.
```

This program creates a one-line text object which contains:

```
hello, world!
```

The routine **Free_text** removes a text object from memory. This is a useful operation, because text objects take up space on the Pascal heap (and the larger the object, the more space it takes up). It is therefore good practice to use **Free_text** to release the space occupied by a text object when you have finished with it. However, *be warned* that if you delete a text object, and then try to use it, chaos will result (it is quite likely that the machine will crash). *Caveat* user!

) *Procedure Free_text*

Routine name: **Free_text**
Kept in unit: **Cit_core**
Purpose: Destroy a text object

The unit name **Cit_core** must appear in the uses clause at the head of your program.

Signature:

```
procedure Free_text (var t : Textptr);
```

Input parameters for Free_text:
Call: **Free_text (t)**
t: Text object to be destroyed

Output parameters for Free_text:
t is set to NIL (the empty text object).

) *8.2.3   The number of lines in a text object*

Let us now introduce the function **Length_text**, which returns the number of lines in a text object.

) *8.2.4   Working example of Length_text*

Filename: **CHAP8\8-2LENGT.PAS**
To run the working example see *Guide to running the software.*

) *Function Length_text*

Routine name: **Length_text**
Kept in unit: **Cit_text**
Purpose: Returns number of lines in a text object
The unit name **Cit_text** must appear in the uses clause at the head of your program.

Signature:

```
function Length_text (t : Textptr) : integer;
```

Input parameters for Length_text:
Call: **Length_text (t)**
t: Text object

Output parameters for Length_text:
Length_text returns the number of lines in the text object.
   Length_text returns 0 if t is the empty text object.
   You will remember that **New_text** returned a text object consisting of one line; clearly, calling **Length_text** on this object should return one. We can check this with the following program:

```
program check_length;
  uses Cit_core, Cit_text;
var t : Textptr;
begin
  t := New_text ('hello, world!');
  writeln (Length_text (t), ' should be 1.')
end.
```

This produces the output:

    1 should be 1.

as we would hope.

⟩ *8.2.5   Keeping track of lines of text*

The operations on text objects set two variables defined in **Cit_text**. **Current_text** points to what we might regard as the 'current line' in the text object; for example, **Append_text** sets **Current_text** to the line which has just been added to the text object. **Current_text** is set to **NIL** if there is no current line within the text object; in this case, the Boolean **End_of_text** is set to **true**. Otherwise, **End_of_text** is set to **false**.

  The routine **Reset_text** merely resets the variables **Current_text** and **End_of_text**.

⟩ *Procedure Reset_text*

Routine name: **Reset_text**
Kept in unit: **Cit_text**
Purpose: Reset **Current_text** and **End_of_text**

The unit name **Cit_text** must appear in the uses clause at the head of your program.

Signature:

    procedure Reset_text (t : Textptr);

Input parameters for Reset_text:
Call: **Reset_text (t);**
t: a text object

Current_text is set to t. **End_of_text** is set **true** if t is **NIL** (the empty text object).

  If t is a text object, the Pascal statement:

    Reset_text (t);

will set **Current_text** equal to t. If t is **NIL** (the empty text object), **End_of_text** will be **true**; otherwise, **End_of_text** will be **false**.

⟩ *8.2.6   Appending lines of text*

Append_text is a simple routine for modifying text objects; it appends the given string to the end of the text object.

⟩ *8.2.7   Working example of Append_text*

Filename: **CHAP8\8-2APPEN.PAS**
To run the working example see *Guide to running the software.*

⟩ *Procedure Append_text*

Routine name: **Append_text**
Kept in unit: **Cit_text**
Purpose: Append a string to a text object

The unit name **Cit_text** must appear in the uses clause at the head of your program.

Signature:

```
procedure Append_text (var t : Textptr;
                            s     : string);
```

Input parameters for Append_text:

Call: **Append_text (t, s)**
t: Text object
s: String to be appended

Output parameters for Append_text:

t is set to the new text object, with the string as its new last line.

As a simple example, the following program creates a text object of two lines:

```
program two_lines;
  uses Cit_core, Cit_text;
var
  t : Textptr;
begin
  t := New_text ('hello, world!');
  Append_text (t, 'what a lovely day!')
end.
```

The text object **t** now contains the two lines shown below:

```
hello, world!
what a lovely day
```

Append_text sets **Current_text** to the line just appended, and **End_of_text** to **false** (since **Current_text** is not the empty text object).

⟩ *8.2.8   Concatenating text objects*

The general procedure **Concat_text** joins two text objects together; as each text object may contain a number of lines, text objects of arbitrary length may easily be assembled in this way. It is important to note that neither text may be any portion of the other, or chaos will result (more accurately, an infinite text will be created). Unfortunately, it is not possible for **Concat_text** to check for this without using a lot of computer time; so, once again, *caveat* user!

⟩ *8.2.9   Working example of Concat_text*

Filename: **CHAP8\8-2CONCA.PAS**
To run the working example see *Guide to running the software.*

⟩ *Procedure Concat_text*

Routine name: **Concat_text**
Kept in unit: **Cit_text**
Purpose: Concatenate two text objects

The unit name **Cit_text** must appear in the uses clause at the head of your program.

Signature:

```
procedure Concat_text (var t : Textptr;
                       t1     : Textptr);
```

Input parameters for Concat_text:
Call: **Concat_text (t, t1)**
t, t1:   Text objects

Output parameters for Concat_text:

t is set to the new text object, containing the original contents of t, followed by those of t1.

Chaos will result if t is any part of t1, or t1 any part of t; but it could take a very long time for Concat_text to check for this.

Here is an alternative program to produce the two-line text we made earlier with Append_text:

```
program two_lines;
  uses Cit_core, Cit_text;
var
  t1, t2 : Textptr;
begin
  t1 := New_text ('hello, world!');
  t2 := New_text ('what a lovely day!');
  Concat_text (t1, t2)
end.
```

The first statement sets t1 to the one-line text hello world!; the second sets t2 to the one-line text what a lovely day!. The call of Concat_text now sets t1 to be the new two-line text object, containing the text:

```
hello, world!
what a lovely day!
```

⟩ 8.2.10   Inserting lines of text

The last routine for adding lines to a text object is called Insert_text. It will add a line anywhere within a text object.

⟩ 8.2.11   Working example of Insert_text

Filename: CHAP8\8-2INSER.PAS
To run the working example see *Guide to running the software.*

⟩ Procedure Insert_text

Routine name:	Insert_text
Kept in unit:	Cit_text
Purpose:	Insert a string into a text object before a given line

The unit name **Cit_text** must appear in the uses clause at the head of your program.

Signature:

```
procedure Insert_text (var t : Textptr;
                           line  : integer;
                           s     : string);
```

Input parameters for Insert_text:

Call:	Insert_text (t, line, s)
t:	Text object
line:	Line number before which string is to be inserted
s:	String to be inserted

Output parameters for Insert_text:

t is set to the new text object, with the string inserted. **Current_text** is set to point to the new line; **End_of_text** is set to **false** (as there is at least one line remaining in the text object).

As a simple example, the Pascal statement:

```
Insert_text (t, 2, 'I''m about to say something:');
```

would insert **I'm about to say something:** *before* the second line of the text object **t**. If **t** contained the two-line text we saw earlier, the statement above would result in **t** containing:

```
hello, world!
I'm about to say something:
what a lovely day!
```

⟩ *8.2.12   Deleting lines of text*

All the procedures we have seen so far add lines to text objects; **Delete_text**, as its name implies, deletes lines within a text object. It deletes a number of lines, starting at a particular point.

⟩ *8.2.13   Working example of Delete_text*

Filename: **CHAP8\8-2DELET.PAS**

To run the working example see *Guide to running the software.*

⟩ *Procedure Delete_text*

Routine name: `Delete_text`
Kept in unit: `Cit_text`
Purpose: Delete lines within a text object

The unit name `Cit_text` must appear in the uses clause at the head of your program.

Signature:

```
procedure Delete_text (var t  : Textptr;
                       line1,
                       nlines : integer);
```

Input parameters for `Delete_text`:

Call: `Delete_text (t, line1, nlines)`
`t`:   Text object
`line1`:   First line to be deleted
`nlines`:   Number of lines to be deleted

Output parameters for `Delete_text`:

`t` is set to the text object, with the requested lines deleted. If this results in all the lines within `t` being deleted, `End_of_text` is set to **true**, otherwise **false**.

For example, if `t` is a text object, then the Pascal statement:

```
Delete_text (t, line, nlines);
```

deletes `nlines` lines, beginning with the one numbered `line`.

⟩ *8.2.14 Save and loading text to and from disc files*

Text objects can contain an arbitrary number of lines, and may be the result of extensive computation; the Toolkit therefore provides simple facilities for saving them to files, and recovering them later.

⟩ *8.2.15 Working example of Save_text*

Filename: **CHAP8\8-2SAVE.PAS**

To run the working example see *Guide to running the software*.

⟩ *Procedure Save_text*

Routine name: `Save_text`
Kept in unit: `Cit_text`
Purpose: Save a text object to a file

The unit name `Cit_text` must appear in the uses clause at the head of your program.

Signature:

```
procedure Save_text (t        : Textptr;
                     filename : Linestring;
                     tag      : Linestring);
```

Input parameters for `Save_text`:

Call: `Save_text (t, filename, tag)`
`t`:   Text object
`filename`: Filename to save the text object to
`tag`:   'tag' given to the text object

Error indicators for `Save_text`:

`Errorflag` is set `true` if a file error occurs.

The procedure `Save_text` saves a text object into a named file. Each file containing a text object is given a 'tag' on its first line; a text object saved with one particular tag can only be reloaded if that tag is specified when the file is loaded. This helps to prevent text objects being initialised from arbitrary files.

Imagine that we have a text object in variable `t`, and wish to save it to the file `MYTEXT.TXT`. `Save_text` takes three parameters: the text object, the file name, and the 'tag' for the file; so:

```
Save_text (t, 'MYTEXT.TXT', 'my text object');
```

would save the text object `t` in the file called `MYTEXT.TXT`, with a tag of `my text object`. In fact, the `.TXT` extension to the file name is optional; `Save_text` will append `.TXT` to a file name without an extension. We might just as well have used:

```
Save_text (t, 'MYTEXT', 'my text object');
```

The complementary operation **Load_text** loads a text object from a given file name, with a given tag.

⟩ *8.2.16 Working example of Load_text*

Filename: **CHAP8\8-2LOAD.PAS**
To run the working example see *Guide to running the software.*

⟩ *Procedure Load_text*

Routine name: **Load_text**
Kept in unit: **Cit_text**
Purpose: Load a text object from a file

The unit name **Cit_text** must appear in the uses clause at the head of your program.

Signature:

```
procedure Load_text (var t     : Textptr;
                          filename : string;
                          tag      : string);
```

Input parameters for Load_text:

Call: **Load_text (t, filename, tag)**
t: Text object
filename: Filename from which to load the text object
tag: Tag with which the text object was saved

Output parameters for Load_text:

t is set to the loaded text object.

Error indicators for Load_text:

If an error occurs (for example, the file cannot be found, or the tag does not match), **Errorflag** is set **true**, and **t** is set to **NIL**, the null text object.

For example, if the file **MYTEXT.TXT** contains a text object saved with the tag **my text object**, we can reload it by using:

```
Load_text (t, 'MYTEXT.TXT', 'my text object');
```

Once again, the .TXT extension on the file name is optional; MYTEXT would have done just as well.

⟩ 8.2.17 *Writing characters to a line of text*

All the subprograms we have seen so far are able to change text objects one line at a time; we now deal with a procedure which can make changes within a given line of a text object.

Write_text overwrites a particular line of a text object, starting at a given column, with the contents of a string.

⟩ 8.2.18 *Working example of Write_text*

Filename: CHAP8\8-2WRITE.PAS
To run the working example see *Guide to running the software.*

⟩ *Procedure Write_text*

Routine name: Write_text
Kept in unit: Cit_text
Purpose: Change one line within a text object

The unit name Cit_text must appear in the uses clause at the head of your program.

Signature:

```
procedure Write_text (var t : Textptr;
                          col  : integer;
                          line : integer;
                          s    : string);
```

Input parameters for Write_text:

Call: Write_text (t, col, line, s)
t: Text object
col: Column at which replacement is to start
line: Line to be affected
s: String to be written

Output parameters for Write_text:

t is set to the new text object.

Imagine that there are n characters in the string; then `Write_text` deletes n characters in the line, starting with the one in column `col`; it then replaces them by the contents of the string. The first column of the line is numbered 1.

As a simple example, imagine that t contains one line:

> `The cat sat on the mat.`

The Pascal statement:

> `Write_text (t, 5, 1, 'dog');`

would delete three characters (the string `'dog'` contains three characters), beginning at column five, and replace them by **dog**. Now line one of the original text object contains the three characters **cat** starting at column five; so the end result would be that t would contain:

> `The dog sat on the mat.`

If `col` is zero or negative, `Write_text` moves along by $|col|$ columns; thus, if the `Write_text` statement above is immediately followed by:

> `Write_text (t, 0, 1, ' chewed');`

t will contain:

> `The dog chewed the mat.`

⟩ *8.2.19   Reading characters from a line of text*

The procedure `Read_text` allows you to read part of a particular line within a text object.

⟩ *Procedure Read_text*

Routine name:	`Read_text`
Kept in unit:	`Cit_text`
Purpose:	Extract a number of characters from one line of a text object

The unit name `Cit_text` must appear in the uses clause at the head of your program.

Signature:

```
procedure Read_text (t     : Textptr;
                      col   : integer;
                      line  : integer;
                      var s : string);
```

Input parameters for Read_text:

Call:	Read_text (t, col, line, s)
t:	Text object
col:	First column from which text is taken
line:	Line number from which text is taken

Output parameters for Read_text:

s is set to the contents of line line of the text object, beginning at column col.

Consider our earlier example, where t was a text object containing the one line:

```
The cat sat on the mat.
```

Read_text has the same parameters as Write_text; if s is a Pascal string variable,

```
Read_text (t, 5, 1, s);
```

sets s to the first line of t, starting with column five. After this Pascal statement, s will contain:

```
cat sat on the mat.
```

Note that Read_text sets s to the *entire* remainder of the line; the number of characters read does not depend on the length of s (this is in contrast to Write_text, where the number of characters modified depends on the length of the string).

Here is a simple program which uses the text routines we have been describing to print the contents of a text object:

```
program show_readtext;
uses Cit_core, Cit_text;
var i,len: integer; t: Textptr; s: string;
begin
  { code to initialise t }
```

```
    len:=Length_text(t);
    for i:=1 to len do begin
      Read_text(t,i,1,s);
      Writeln(s)
    end;
  end.
```

⟩ 8.2.20 Shifting the base line

The procedure Shift_text is defined in the unit Cit_core; it is used internally by Cit_text and Cit_edit, and has a highly specialised function: to move the reference point through a text object.

Normally, the pointer variable (of type Textptr) points to the beginning of the text object. The lines of text are then chained into a list that connects all the lines in sequential order. If you want to find a particular line, you must search through from the beginning. However, the routine Shift_text can be used to move the start of the chain forward a given number of lines. This has the effect of reducing the time needed to scan through the text for particular lines.

⟩ Procedure Shift_text

Routine name: Shift_text
Kept in unit: Cit_core
Purpose: Shift the base line for the text list

The unit name Cit_core must appear in the uses clause at the head of your program.

Signature:

```
      procedure Shift_text (t        : Textptr;
                            VAR line : integer);
```

Input parameters for Shift_text:
Call: Shift_text (t, line)
t: Text object
line: Desired number of lines to shift

Output parameters for Shift_text:
line: Actual number of lines shifted

The routine steps through the chain of text lines. For each step it reduces the value of the parameter line by one. If there are enough lines in the text for the shift, then the parameter line returns the value 1. Otherwise, it returns a value which is 1 plus the shortfall.

There is an important application which can be used as an example. Suppose you wish to list a text object to the screen. The following code will list an existing text object t:

```
program show_shift;
uses Cit_core, Cit_text;
var line: integer; t: Textptr; s: string;
begin
  { insert code here to set t }

  line:=1;
  while line=1 do begin
    Read_text (t,1,1,s);
    Writeln (s);
    line:=2;
    Shift_text (t,line);
  end
end.
```

This code performs the same function as the example code for Read_text; but note that the code above modifies t.

) Chapter 9

) Formatting Values and Evaluating Expressions

This chapter describes sets of facilities for the conversion of numerical values into strings of characters, and for the conversion of strings into numerical values.

In the Turbo Pascal environment it is quite possible to convert values into strings and to output values directly to the screen. But the Toolkit has a rich set of display routines which make demands better met by its own formatting routines. These routines are contained in the unit **Cit_form** and provide facilities for formatting single values and arrays of values in a manner suited to the rest of the Toolkit.

By way of contrast there is no natural way in Turbo Pascal of converting general string expressions to values; what we need is a string evaluator. The Toolkit provides one in the unit **Cit_eval**. It is accessible in a variety of ways, and the routines can be arranged to give repeated fast evaluation of string expressions.

) 9.1   Unit Cit_form

An important part of numerical programming is presenting the results in a way which is easy to understand. Standard Pascal allows you to control the format in which values are printed by means of optional arguments to **Write** and **Writeln**.

However, these procedures are not always general enough. If you try to display a value in a field which is too small, **Writeln** outputs the value in full, ignoring the field width. Thus, for example:

```
Writeln (12345 : 3);
```

(which attempts to write the value 12345 in three characters) will print

```
12345
```

and produce five characters of output. This is bad enough in a table, because it spoils the layout; but is still worse if you are displaying the value in a window, as the window may be too small to hold the extra characters.

Using the Toolkit routine `Format_real` to format 12345 into a string of three characters would produce:

    1E4

which is the best that can be done within the small space available. If the value really is too large to be displayed (for example, you are trying to format 12345 into a string of two characters), the Toolkit routines produce a null string. This does not cause any confusion, as the failure of a number to appear on a display is a clear sign that something has gone wrong; moreover, a null string is guaranteed to fit any space, however small.

The Toolkit routines format values in three styles:

- **Fixed-point** formatting formats the value without an exponent, but with a specified number of figures after the decimal point. Obviously, only values fairly close to one can be formatted in this manner; if you try to format 10^{30} in fixed-point format, the integer part consists of a one followed by thirty zeroes, which is not particularly clear!

- **Floating-point** formatting produces a formatted value with an exponent. This is much clearer for values very different from one; formatting 10^{30} might produce:

    1.0E30

which is much better than a one followed by thirty zeroes! On the other hand, the number two is much better displayed in fixed-point format.

- **General** formatting tries to decide for itself whether to format numbers in fixed-point or floating-point style. If there is room to display the value in a fixed-point style, general formatting uses that; the number of places after the decimal point depends on the amount of space remaining in the field. Otherwise, values are displayed in floating-point style, with the number of places displayed depending on the field width.

The Pascal procedures `Write` and `Writeln` output values to files; the Toolkit routines, by contrast, format them to strings (like the Turbo

Pascal function Str). These strings can easily be modified or passed on to other routines, and it is easy to use **Write** or **Writeln** to output them to files, if so required.

The Toolkit routines allow you to format either individual values, or matrices of values, in one operation. The latter is useful if you have created or calculated a matrix of values (perhaps using the Toolkit matrix routines), and wish to display it without having to do much programming yourself.

⟩ *9.1.1   Routine* Format_real

The procedure **Format_real** is the most general and 'automatic' of those in **Cit_form**. It displays a numeric value in a field of given width, and decides for itself whether or not to display an exponent, and the number of figures after the decimal point, according to the value and the width available. It displays the value in fixed-point format, if there is room, selecting the number of decimal places displayed appropriately; otherwise, it uses floating-point format.

The simplest way of becoming acquainted with **Format_real** is to use it yourself to format values (see the 'Working example' mentioned below). The working example program requests three things from you: the field width in which you want the number formatted (if this is zero or negative, the program stops); the value to format; and the sign required for positive numbers (see below). The program then prints the output from **Format_real**, and then prompts you again for new values.

This sign for positive numbers is given by a single character, with the following meaning:

Character	Positive numbers preceded by
'_'	nothing
'+'	a plus sign
' '	at least one space

Positive numbers can thus appear with no sign, with a plus sign, or with a space instead of a sign. This last option enables both positive and negative numbers to be displayed in a table, without the signs of negative numbers appearing in the same column as the leading digits of positive numbers.

⟩ *9.1.2   Working example of general formatting*

Filename: `CHAP9\9-1REAL.PAS`
To run the working example see *Guide to running the software.*

⟩ *Procedure Format_real*

Routine name: `Format_real`
Kept in unit: `Cit_form`
Purpose: Format a numeric value to fit a given field width, using
                   fixed-point format if possible, floating-point format
                   otherwise.

The unit name `Cit_form` must appear in the uses clause at the head of
your program.

Signature:

```
procedure Format_real (rval      : Citreal;
                       field     : integer;
                       sign      : char;
                       var strng : Linestring);
```

Input parameters for `Format_real`:

Call: `Format_real (rval, field, sign, strng)`
`rval`:   Value to be formatted
`field`:   Field width
`sign`:   Sign for positive numbers

Output parameters for `Format_real`:

`strng` contains the result (the characters representing the value).

Error indicators for `Format_real`:

If the field width is too small to display the value, **strng** is set to the
null string and **Errorflag** is set **true**.

⟩ *9.1.3   Working example of general formatting*

Filename: `CHAP9\9-1FLOAT.PAS`
To run the working example see *Guide to running the software.*

⟩ *9.1.4   Routine* Format_float

Format_float formats a value in floating-point format. The exponent is displayed as an E, followed by the exponent in minimum width; thus 10^{30} would have an exponent of:

   E30

whereas 0.1 would have an exponent of:

   E-1

Format_float always puts one, non-zero, digit before the decimal point; one might be formatted as:

   1.0E0

rather than 0.1E1, as in some other programming languages.

  The simplest way of becoming acquainted with Format_float is to use it yourself to format values. The working example program requests three things from you: the field width in which you want the number formatted (if this is zero or negative, the program stops); the value to format; and the sign required for positive numbers (as for Format_real). The program then prints the output from Format_float, and then prompts you again for new values.

⟩ *Procedure Format_float*

Routine name: Format_float
Kept in unit: Cit_form
Purpose: Format a numeric value to fit a given field width, using floating-point format

The unit name Cit_form must appear in the uses clause at the head of your program.

Signature:

```
procedure Format_float (rval    : Citreal;
                        field    : integer;
                        sign     : char;
                        var strng : Linestring);
```

Input parameters for Format_float:

Call:	Format_float (rval, field, sign, strng)
rval:	Value to be formatted
field:	Field width
sign:	Sign for positive numbers

Output parameters for Format_float:

strng contains the result (the characters representing the value).

Error indicators for Format_float:

If the field width is too small to display the value, strng is set to the null string and Errorflag is set true.

⟩ *9.1.5   Routine* Format_fixpt

Format_fixpt formats values in fixed-point format. Because fixed-point format is used most commonly for tables and matrices, and these have various requirements on the way the values are formatted, Format_fixpt has more parameters, and more flexibility, than the other routines in Cit_form.

Format_fixpt formats a value in a given field width, with a given number of figures after the decimal point. You can control whether a sign appears before positive numbers (as in the previous routines), and, additionally, whether the value is formatted in minimum width, or left- or right-justified within the given field.

The simplest way of becoming acquainted with Format_fixpt is to use it yourself to format values. The working example program requests five things from you:

- the field width in which you want the number formatted (if this is zero or negative, the program stops)
- the number of digits after the decimal point
- a character representing the alignment:
  c for compact formatting
  l to left-justify the value within the field
  r to right-justify the value within the field
- the value to format
- the sign required for positive numbers (as for Format_real). The program then prints the output from Format_fixpt, and then prompts you again for new values.

The alignment is passed to **Format_fixpt** by one of three scalar values:

Value	Meaning
COMPACT	produce the value in minimum width (perhaps smaller than field)
LEFT	place the value at the left-hand edge of the field
RIGHT	place the value at the right-hand edge of the field

) *Procedure Format_fixpt*

Routine name:	**Format_fixpt**
Kept in unit:	**Cit_form**
Purpose:	Format a numeric value to fit a given field width, using fixed-point format

The unit name **Cit_form** must appear in the uses clause at the head of your program.

Signature:

```
procedure Format_fixpt(rval        : Citreal;
                       field, ndec : integer;
                       alignment   : Alignment_type;
                       sign        : char;
                       var strng   : Linestring);
```

Input parameters for Format_fixpt:

Call:	**Format_fixpt (r, f, nd, align, s, strng)**
r:	Value to be formatted
f:	Field width
nd:	Number of digits after the decimal point
align:	Alignment of the value within the field
s:	Sign for positive numbers

Output parameters for Format_fixpt:

strng contains the result (the characters representing the value).

Error indicators for Format_fixpt:

If the field width is too small to display the value, **strng** is set to the null string and **Errorflag** is set **true**.

〉 *9.1.6   Formatting matrices*

Cit_form also contains three procedures which format matrices:

- **Format_array_real** which formats matrix elements in the style of **Format_real**
- **Format_array_float** which formats matrix elements in the style of **Format_float**
- **Format_array_fixpt** which formats matrix elements in the style of **Format_fixpt**

The matrices, which do not have to be square, are contained in two-dimensional arrays. The matrix-formatting procedures format each element using the corresponding value-formatting procedure (for example, **Format_array_real** formats each matrix element with **Format_real**). Positive numbers are preceded by a space, rather than a sign (so that the first digits of positive and negative values in a column line up), and values are right-justified within each field.

The matrices are formatted onto several lines (the number of lines being the number of rows in the matrix). Because strings can only hold a restricted number of characters, and matrices can be very large, the matrix-formatting procedures return their results in text objects (which are held in variables of type **Textptr**) rather than as strings. Text objects are described under unit **Cit_text**, in Chapter 8.

〉 *Procedure Format_array_real*

Routine name:	**Format_array_real**
Kept in unit:	**Cit_form**
Purpose:	Format a matrix of numeric values; format each to fit a given field width, using fixed-point format if possible, floating-point format otherwise

The unit name **Cit_form** must appear in the uses clause at the head of your program.

Signature:

```
procedure Format_array_real(a           : Arrayptr;
                            m, n, n0, f : integer;
                            var t       : Textptr);
```

Input parameters for Format_array_real:

Call:	Format_array_real (@a, m, n, n0, f, t)
a:	Matrix to be formatted
m:	Number of rows in matrix
n:	Number of columns in matrix
n0:	Size of first dimension of matrix
f:	Field width

Output parameters for Format_array_real:

t contains the result (the text object representing the value).

Error indicators for Format_array_real:

If the field width is too small to display the value, the corresponding entry in t is set to the null string and **Errorflag** is set **true**.

⟩ *Procedure Format_array_ float*

Routine name:	Format_array_float
Kept in unit:	Cit_form
Purpose:	Format a matrix of numeric values, using floating-point format

The unit name **Cit_form** must appear in the uses clause at the head of your program.

Signature:

```
procedure Format_array_float(a           : Arrayptr;
                             m, n, n0, f : integer;
                             var t       : Textptr);
```

Input parameters for Format_array_float:

Call:	Format_array_float (@a, m, n, n0, f, t)
a:	Matrix to be formatted
m:	Number of rows in matrix
n:	Number of columns in matrix
n0:	Size of first dimension of matrix
f:	Field width

Output parameters for Format_array_float:

t contains the result (the text object representing the value).

Error indicators for Format_array_float:

If the field width is too small to display the value, the corresponding entry in t is set to the null string and **Errorflag** is set **true**.

) *Procedure Format_array_ fixpt*

Routine name:	**Format_array_fixpt**
Kept in unit:	**Cit_form**
Purpose:	Format a matrix of numeric values; format each to fit a given field width, using fixed-point format

The unit name **Cit_form** must appear in the uses clause at the head of your program.

Signature:

```
procedure Format_array_fixpt(a          : Arrayptr;
                        m, n, n0,
                        f, nd      : integer;
                        var t      : Textptr);
```

Input parameters for Format_array_fixpt:

Call:	Format_array_fixpt (@a, m, n, n0, f, nd, t)
a:	Matrix to be formatted
m:	Number of rows in matrix
n:	Number of columns in matrix
n0:	Size of first dimension of matrix
f:	Field width
nd:	Number of digits after decimal point

Output parameters for Format_array_fixpt:

t contains the result (the text object representing the value).

Error indicators for Format_array_fixpt:

If the field width is too small to display the value, the corresponding entry in t is set to the null string and **Errorflag** is set **true**.

⟩ 9.2 Unit Cit_eval

You may be surprised to find that the Toolkit contains a unit which enables you to evaluate expressions; you can do this, after all, perfectly well in Pascal. However, Pascal can only evaluate those expressions which are known when a program is compiled; there are many instances where this might not be the case.

For example, imagine a simple graph-drawing program, which plots a function over some range. Ideally, we would like the program to be able to ask the user to input the function to plot; but then the function we wish to evaluate depends on what the user types in!

The Toolkit contains code which can evaluate expressions like this, which are only known when the program is run.

The evaluator can, optionally, be told of the existence of variables; thus you can evaluate expressions like:

```
2 * sin(x) * cos(x)
```

where **x** represents a value contained in a Pascal variable. This makes it very straightforward to specify functions or plot graphs.

There are many different ways of using the evaluator. Some are very straightforward; but the more advanced methods give you more power, speed and flexibility. As usual in this Toolkit manual, we begin with simple examples, and work up to more complex ones.

Let us add a note about the evaluator's checking for numerical overflow. By default, it does not check for numerical overflow, which means that some operations, such as multiplication, may produce a run-time error from Turbo Pascal. This is normally satisfactory, as something has clearly gone wrong if the calculation overflows.

However, if you are intending to produce a robust program (say for student use), this behaviour is inconvenient. By recompiling the source of the Toolkit, and specifying the conditional define **CHECKING**, you can make the evaluator test for, and report, numeric overflow. This slows the evaluator down, and reduces the range of values it can produce; but is useful for producing robust programs. For information on setting conditional defines, and recompiling the Toolkit, see Chapter 1.

⟩ 9.2.1 *Expressions to be evaluated*

The evaluator accepts expressions in character form, either as strings or as text objects (which are described in Chapter 8, under unit Cit_text).

These expressions may use numbers and operators, just as in Pascal, and (optionally) variable values as well. The evaluator will accept any expression which is legal in Pascal; in addition, it allows a few extensions (the most important being that ^ may be used as an exponentiation operator, so that **x^y** returns the value x^y).

All the usual Pascal functions for real values, like **sin**, are available, along with a few which aren't (**arcsin** and **tan**, for example). If you are interested in the exact meaning of expressions (their 'grammar', to use the technical term), you should consult the last portion of this section.

⟩ *9.2.2   Routine* **Eval**

The function **Eval** is the simplest way of using the evaluator. It is passed an expression as a variable of type **string**, and a Boolean, which we shall set to **false** for the minute*. **Eval** returns the result of evaluating the expression.

The simplest program to use **Eval** is the following:

```
program show_eval;
  uses Cit_eval;
var expr : string;
begin
  Write ('Please input your expression: ');
  Readln (expr);
  Writeln ('This evaluates to ', Eval (expr, false))
end.
```

When you run this program, it will request an expression, and then print the result of evaluating it.

⟩ *9.2.3   Working example of Eval*

Filename: **CHAP9\9-2EVAL.PAS**
To run the working example see *Guide to running the software*.

⟩ *Function Eval*

Routine name: **Eval**

---

* The Boolean controls whether variables may be used; we shall discuss this later.

Kept in unit: Cit_eval
Purpose: Evaluate an expression held in a string

The unit name Cit_eval must appear in the uses clause at the head of your program.

Signature:

```
function  Eval(expr    : string
                can_vary : boolean): Citreal;
```

Input parameters for Eval:

Call: Eval (expr, can_vary)
expr: String containing expression to be evaluated
can_vary: Boolean giving whether variables can be used

Output parameters for Eval:

The function returns the result of the evaluation.

Error indicators for Eval:

The error procedure **Seterror** is called if an error is found whilst evaluating the string; this, amongst other things, sets the global variable **Errorflag**.

) 9.2.4   *Compiling expressions*

Eval works very simply and effectively, but has one drawback: it is relatively slow. This is obviously not a problem if an expression needs to be evaluated only once; but it is more serious if an expression needs to be evaluated many times (for example, the expression for a function may need to be evaluated many times if it is to be plotted).

The Toolkit evaluator has an option which enables expressions to be worked out much more quickly. Before any evaluation is done, the expression is first 'compiled'. This compilation involves converting the expression from a string of characters into an internal form*. This internal form is much quicker to evaluate than the original; for instance, sequences of digits have been converted into the values they represent. Compilation is a relatively slow process; but compiled expressions may then be evaluated far more quickly. Thus, if an expression is to be evaluated many times (if, for instance, it is used to define a graph, and is

---

* In fact, a simple parse tree.

evaluated at a large number of points), using compiled expressions will save a lot of time.

Expression compiling can be likened to compiling Pascal; a Pascal program must be compiled (a fairly slow process) before it can be run. However, once compiled, it can be run many times without recompilation.

Compiled expressions are held in variables of type `Expptr`; this type is declared in the unit `Cit_core`. `Compile_string` is passed a string and a Boolean, just like `Eval`; it returns the compiled expression in a third parameter of type `Expptr`. For example:

```
program show_compile;
  uses Cit_core, Cit_eval;
var expr : string; xp : Expptr;
begin
  Write ('please input your expression: ');
  Readln (expr);
  Compile_string (expr, false, xp)
end.
```

Obviously this is not a very useful program, but it does show how easy it is to compile an expression. `Compile_string`'s Boolean argument has the same meaning as for `Eval` (and, just like `Eval`, we shall set the Boolean false for the minute).

) 9.2.5   *Working example of Compile_string*

Filename: `CHAP9\9-2COMPI.PAS`
To run the working example see *Guide to running the software.*

) *Procedure Compile_string*

Routine name: `Compile_string`
Kept in unit: `Cit_eval`
Purpose: Compile an expression held in a string

The unit name `Cit_eval` must appear in the uses clause at the head of your program.

Signature:

```
procedure Compile_string(expr    : string;
                         can_vary: boolean;
                         var expp: Expptr);
```

Input parameters for Compile_string**:**

Call:	Compile_string (expr, can_vary, expp)
expr:	String containing expression to be evaluated
can_vary:	Boolean giving whether variables can be used

Output parameters for Compile_string**:**

expp is set to the internal form of the expression compiled, which can subsequently be evaluated with the function Evaluate.

Error indicators for Compile_string**:**

The error procedure Seterror is called if an error is found whilst compiling the expression; this, amongst other things, sets the global variable Errorflag.

 If a name is found to be undeclared, the global variable Error_name is set to the offending name; this makes it easier to produce messages like "unknown name 'x'".

) *9.2.6   Routine Compile_text*

If you have to compile a very long expression (for example, a large polynomial), you may find that a Turbo Pascal string is not long enough to hold all the characters required. In this case, you can use the Toolkit's text objects, which are held in variables of type Textptr (described under unit Cit_text), to contain an expression of arbitrary length. Strings held in Textptr variables are compiled using Compile_text rather than Compile_string, but the procedures are otherwise identical.

) *Procedure Compile_text*

Routine name:	Compile_text
Kept in unit:	Cit_eval
Purpose:	Compile an expression held in a Textptr

The unit name Cit_eval must appear in the uses clause at the head of your program.

Signature:

```
procedure Compile_text(expr   : Textptr;
                       can_vary: boolean;
                       var expp: Expptr);
```

Input parameters for Compile_text:

Call: `Compile_text (expr, can_vary, expp)`
expr: Text pointer containing expression to be evaluated
can_vary: Boolean giving whether variables can be used

Output parameters for Compile_text:

expp is set to a representation of the expression compiled, which can subsequently be evaluated with the function **Evaluate**.

Error indicators for Compile_text:

The error procedure **Seterror** is called if an error is found whilst compiling the expression; this, amongst other things, sets the global variable **Errorflag**.

⟩ *9.2.7   Evaluating compiled expressions*

Once an expression has been compiled, it can be evaluated by calling the function **Evaluate**. **Evaluate** returns the result of evaluating the compiled expression. The compiled expression is unchanged by the call of **Evaluate**; **Evaluate** may therefore be called many times for a given compiled expression. **Evaluate** returns a value of type **Citreal**.

⟩ *9.2.8   Working example of Evaluate*

Filename: **CHAP9\9-2COMPI.PAS**
To run the working example see *Guide to running the software.*

⟩ *Function Evaluate*

Routine name: **Evaluate**
Kept in unit: **Cit_eval**
Purpose: Evaluate a compiled expression

The unit name **Cit_eval** must appear in the uses clause at the head of your program.

Signature:

```
function Evaluate (expp : Expptr) : Citreal;
```

Input parameters for Evaluate:

Call: Evaluate (expp)

expp: Variable of type Expptr

Output parameters for Evaluate:

Evaluate returns a value of type Citreal, which is the result of evaluating the expression.

Error indicators for Evaluate:

The error procedure Seterror is called if an error is found whilst evaluating the string; this, amongst other things, sets the global variable Errorflag.

⟩ 9.2.9   Declaring names

The evaluator stores the meanings of names like the predefined functions sin and sqrt in its **dictionary**. You may, if you wish, insert names of your own choosing into the dictionary; these may represent constants, variables, arrays or functions, as required. Declaring names works regardless of whether you use Eval to evaluate an expression directly, or Compile_string to compile the expression, followed by Evaluate to determine its value.

Let us begin with the simplest case: declaring a constant. Mathematicians normally use e to denote the base of natural logarithms ($e = 2.718\ldots$). To declare a name e containing this value, we use the procedure Declare_constant:

```
Declare_constant ('e', false, Exp (1.0));
```

Exp (1.0) returns the value of e as accurately as the machine can calculate it.

This middle parameter (quietly set to false above) controls whether the case of letters in the name is significant. Consider another declaration, of a constant Max with a value ten:

```
Declare_constant ('Max', false, 10.0);
```

This would declare **Max** so that case was not significant; that is, you could spell it **Max** or **max** or **MAX** (or even **mAx**, if you wished).

By contrast, declaring **Max** using:

```
Declare_constant ('Max', true, 10.0);
```

would declare **Max** so that it would only be recognised if spelled exactly **Max**; none of the other possibilities, like **max**, would be recognised.

⟩ *9.2.10   Working example of Declare_constant*

Filename: **CHAP9\9-2DECLA.PAS**
To run the working example see *Guide to running the software*.

⟩ *Procedure Declare_constant*

Routine name: **Declare_constant**
Kept in unit: **Cit_eval**
Purpose: Declare a name to represent a constant value

The unit name **Cit_eval** must appear in the uses clause at the head of your program.

Signature:

```
procedure Declare_constant (name : Linestring;
                            note : boolean;
                            valu : Citreal);
```

Input parameters for Declare_constant:

Call: **Declare_constant (name, note, valu)**
name: Name to be declared
note: Boolean, giving case significance
valu: Value of the constant

Output parameters for Declare_constant:
None.

Error indicators for Declare_constant:
None.

⟩ *9.2.11   Routine Declare_variable*

The next simplest operation is declaring a name to correspond to the value in a Pascal variable. There is an important point to note here: the name will correspond to the value in the Pascal variable *at the time the expression is evaluated.*

Imagine that an expression involving a variable is evaluated; the corresponding Pascal variable changed; and the expression evaluated again. The two evaluations will, in general, return *different* results, because the variable's value has changed. Let us give a simple example of this:

```
Declare_variable ('x', false, x_variable);
```

where x_variable is a Pascal variable of type Citreal. Note that there is no need for the Pascal variable to have the same name as the declared variable!

Imagine now we execute the following code:

```
x_variable := 10.0;
Writeln (Eval ('x + 1', true));
```

This program fragment will print out 11.0, because the evaluator variable x has the value 10, and we add one to it. Imagine that we now execute the following two statements:

```
x_variable := 11.0;
Writeln (Eval ('x + 1', true));
```

This will now print out 12.0, because x_variable, the Pascal variable corresponding to the evaluator variable x, has the value 11. To repeat: if the Pascal variable used in the call of Declare_variable changes its value, the corresponding evaluator variable changes, too. (If x_variable changes, then so does the evaluator's variable x.)

You will have noted that Eval was called above with a second parameter of true. This second parameter controls whether or not variables are allowed in expressions. You may find this a little strange: why should an expression not permit variables?

But there are cases where an expression should contain only constants. Consider, for example, a graph plotter, using the variables **x** and **y**, say. When we evaluate the expression for the curve, we obviously need to use the variables **x** and **y**. However the expressions for the upper and lower bounds of the axes must be independent of **x** and **y**; in such cases, it is useful to set the Boolean **can_vary** to **false**.

) *Procedure Declare_variable*

Routine name: **Declare_variable**
Kept in unit: **Cit_eval**
Purpose: Declare a name to represent a variable value

The unit name **Cit_eval** must appear in the uses clause at the head of your program.

Signature:

```
procedure Declare_variable (name : Linestring;
                            note : boolean;
                            valu : var Citreal);
```

Input parameters for Declare_variable:

Call: Declare_variable (name, note, valu)
name: Name being declared
note: Boolean, determining whether case matters
valu: Value of the declared name

Output parameters for Declare_variable:
None.

Error indicators for Declare_variable:
None.

) *9.2.12   Routine Declare_array*

Expressions may contain subscripted, as well as normal, variables. Such variables must be followed immediately by a subscript, enclosed in either square or round brackets (for example, **a(3)** or **data [13]**). Subscripted variables may be declared, and associated with values in a Pascal array, by the procedure **Declare_array**. This procedure is also passed the upper and lower *bounds* of the subscript; if the subscript lies outside this range, an error will occur when the expression is evaluated. The lower bound of the declared array must be the same as the lower bound of the associated Pascal array.

) *Procedure Declare_array*

Routine name: `Declare_array`
Kept in unit: `Cit_eval`
Purpose: Declare a name to represent an array

The unit name `Cit_eval` must appear in the uses clause at the head of your program.

Signature:

```
procedure Declare_array   (name   : Linestring;
                           note   : boolean;
                           var ap : Arrayptr;
                           lb,ub  : integer);
```

Input parameters for `Declare_array`:

Call:	`Declare_array (name, note, @arry, lower, upper)`
name:	Name being declared
note:	Boolean, determining whether case matters
arry:	Array containing values for the declared array
lower:	Array's lower bound
upper:	Array's upper bound

Output parameters for `Declare_array`:

None.

Error indicators for `Declare_array`:

None.

) *9.2.13   Routine Declare_function*

The evaluator has built into it a large number of functions, like **sin**, from Turbo Pascal. Suppose, however, that we wish to evaluate expressions like (say) cosec(x) (cosec is not provided by the evaluator). The first step is to provide our own function to evaluate this:

```
{$F+}
function my_cosec (theta : Citreal) : Citreal;
begin
  my_cosec := 1.0 / sin (theta);
end;
{$F-}
```

(This function is merely provided for illustration; it would probably need modification for serious use, especially for values of **theta** close to zero.) Note that the function must have one parameter; that this must be of type **Citreal**; and that it must return a result of type **Citreal**. In addition, note that it must be compiled with the far-calling option on (the purpose of the {$F+} above).

We can now declare the evaluator function **cosec** as follows:

```
Declare_function ('cosec', false, @my_cosec);
```

Note that **Declare_function** is very similar to the other declaration procedures; instead of a value, however, it is passed the address of the Pascal function which computes the result (the **@** operator returns the address of the item following it). Note that you cannot pass a built-in function, like **Sin**, to **Declare_function**.

) *Procedure Declare_ function*

Routine name: **Declare_function**
Kept in unit: **Cit_eval**
Purpose: Declare a name to represent a Pascal function

The unit name **Cit_eval** must appear in the uses clause at the head of your program.

Signature:

```
procedure Declare_function    (name    : Linestring;
                               note    : boolean;
                               fp      : Funcptr);
```

Input parameters for Declare_function:

Call:	Declare_function (name, note, @fn)
name:	Variable of type Linestring, which contains the name being declared
note:	Boolean, determining whether case matters
@fn:	Where fn is a Pascal function of one Citreal variable

Output parameters for Declare_function:

None.

Error indicators for Declare_function:

None.

⟩ *9.2.14   Undeclaring names*

You may find that you want the evaluator to 'forget' certain names; there are two ways of doing this.

If you want the evaluator to forget every name you have declared, and put the dictionary back into its original state, use the procedure:

        Reset_dictionary;

If, however, you merely wish to remove the declaration of one name, use the procedure Undeclare:

        Undeclare ('x');

for example, will remove the declaration of x, regardless of whether x was declared to be a constant, variable, array or function.

⟩ *Procedure Reset_dictionary*

Routine name:	Reset_dictionary
Kept in unit:	Cit_eval
Purpose:	Remove all names declared by user

The unit name Cit_eval must appear in the uses clause at the head of your program.

Signature:

```
        procedure Reset_dictionary;
```

Input parameters for Reset_dictionary:

Call: `Reset_dictionary;`

Output parameters for Reset_dictionary:
None.

Error indicators for Reset_dictionary:
None.

⟩ *Procedure Undeclare*

Routine name: **Undeclare**
Kept in unit: **Cit_eval**
Purpose: Remove a declared name

The unit name **Cit_eval** must appear in the uses clause at the head of your program.

Signature:

```
        procedure Undeclare (name : Linestring);
```

Input parameters for Undeclare:

Call: `Undeclare (name)`
name: Name to be removed

Output parameters for Undeclare:
None.

Error indicators for Undeclare:
None.

⟩ *9.2.15   Freeing expression memory*

When an expression is compiled, the compiled expression is stored on the heap. Whilst only a small amount of memory is used, you may find it worthwhile to free this memory once you have finished with an expression. The way to do this is to call the procedure **Free_tree**:

        Free_tree (expp);

releases the memory occupied by the compiled expression tree stored in **expp**. Once **Free_tree** has been called, any attempt to evaluate **expp** will cause total chaos (quite likely crashing the machine). So call **Free_tree** only if you are sure that the expression is not going to be evaluated again.

⟩ *Procedure Free_tree*

Routine name: **Free_tree**
Kept in unit: **Cit_eval**
Purpose: Release the memory occupied by a compiled expres-
                   sion

The unit name **Cit_eval** must appear in the uses clause at the head of your program.

Signature:

        procedure Free_tree (expp : Expptr);

Input parameters for Free_tree:
Call: **Free_tree (expp)**
expp: Expression to be released

Output parameters for Free_tree:
None.

Error indicators for Free_tree:
None.

⟩ *9.2.16 Evaluating several expressions at once*

There are cases where it is useful to be able to evaluate a number of expressions, and store their results in an array; a typical case would be to evaluate a matrix whose components are expressions, rather than constants.

This may be done by using the procedure **Eval_array**, which performs much the same work as **Eval**, but with two major differences: the expressions are held in a text object (a variable of type **Textptr**), rather than in Turbo Pascal strings (because the matrix of expressions is potentially very large); secondly, the results are returned to an array parameter, rather than returned as the result of the function (functions cannot return arrays in Pascal).

The elements of the matrix are represented in the text object as follows. Each line within the text object represents a row of the matrix, with the first line being row one. Within each line of the text object, the matrix elements are represented by expressions separated by spaces, with the first expression belonging to column one. Obviously, the expressions should not contain spaces. Missing expressions are treated as though they had a value of zero.

) *Procedure Eval_array*

Routine name: **Eval_array**
Kept in unit: **Cit_eval**
Purpose: Evaluate a matrix of expressions

The unit name **Cit_eval** must appear in the uses clause at the head of your program.

Signature:

```
procedure Eval_array(t       : Textptr;
                     can_vary: boolean;
                     a       : Arrayptr;
                     m,n,n0  : integer);
```

Input parameters for Eval_array:

Call: Eval_array (t, can_vary, @a, m, n, n0)
t: Variable of type **Textptr**, which contains the expressions to be evaluated
can_vary: Boolean, giving whether variables can be used
m: Number of rows in the matrix
n: Number of columns in the matrix
n0:   First dimension of **a**

Output parameters for Eval_array:

a: Two-dimensional array of **Citreal**

Error indicators for Eval_array:

If an error occurs whilst evaluating an expression, **Seterror** is called; this sets the global flag **Errorflag**, amongst other things.

⟩ *9.2.17   The grammar of expressions*

The evaluator deals with expressions in almost exactly the same way as Pascal, so any expression given to the formatter will produce the same result as putting that expression into a Pascal program. There are, however, one or two minor changes:

- Characters with ordinal values 128 to 255 are allowed in identifiers (this facilitates putting in your own variables with special names, like π containing the value 3.1415...).
- Primes are allowed in variable names; this means that the evaluator can have variables called things like **f'**, which is pleasant when expressions use derivatives.
- Square brackets means exactly the same as round brackets; either can be used to enclose an array subscript or a function argument.
- As noted earlier, the operator ^ performs exponentiation:

      x ^ y

returns the value x^y. This operator is unusual, because it binds to the right; that is:

      x ^ y ^ z

means x^{y^z}, rather than $(x^y)^z$. This is different from the other operators, which bind to the left, as normal:

      x - y - z

means $(x - y) - z$.

- Numbers may start with a decimal point (Pascal insists that all numbers start with a digit).

⟩ *Predefined functions*

Here is the full list of predefined functions the evaluator knows; these
are identical to those known by Pascal, with a few additions (like **tan**):
- **exp** returns e^x.
- **ln** returns $\log_e x$, $x > 0$.
- **log** returns $\log_{10} x$, $x > 0$.
- **sqrt** returns $\sqrt{x}$, $x \geq 0$.
- **sin** returns $\sin x$.
- **cos** returns $\cos x$.
- **tan** returns $\tan x$.
- **atan** returns $\arctan x$.
- **acos** returns $\arccos x$, $|x| \leq 1$.
- **asin** returns $\arcsin x$, $|x| \leq 1$.
- **cosh** returns $\cosh x$.
- **sinh** returns $\sinh x$.
- **tanh** returns $\tanh x$.
- **abs** returns $|x|$.
- **rnd** returns a random value as follows: if $x = 1$, it returns a real value
  v, with $0 \leq v < 1$; otherwise, for integer x, $1 < x < 32766$, it returns
  an integer, i, with $1 \leq i \leq x$.
- **fnH** is the Heaviside function: if $x < 0$, it returns zero, otherwise 1.

If any of the functions detect an error (for instance, **sqrt** is called with a
negative argument, or the result overflows), an error occurs at evaluation
time.

⟩ Appendix 1

⟩ Unit Cit_core

⟩ A1.1   Introduction

Most of the Toolkit units contain large numbers of useful functions and procedures. **Cit_core** does not contain many of these; its importance lies in the number of fundamental *types* and *constants* it defines, which are central to the Toolkit. For example, **Cit_core** contains the definitions of the Toolkit's floating-point type **Citreal**, and the colour constants (such as **CITBLACK**). The routines which **Cit_core** contains are used widely throughout the Toolkit, but it is fairly unlikely that you would use them directly yourself.

This description of **Cit_core** therefore concentrates on the various types and constants; we leave the routines for later, and then describe them only briefly.

The most important, and pervasive, definition in **Cit_core** is that of the type **Citreal**. The Toolkit uses this type throughout whenever a floating-point type is required, rather than use any of the Turbo Pascal floating-point types directly. This is quite deliberate: it is very easy to change the definition of **Citreal** to the Turbo Pascal type you want, but it would be very tedious (and error-prone) to have to edit the whole Toolkit to change the floating-point type it uses.

If you have purchased the TPU-file version of the Toolkit, the type **Citreal** will be identical to the Turbo Pascal **real** type, which occupies six bytes, holds numbers to approximately eleven digits of accuracy, and has an exponent range from about 10^{-38} to 10^{38}. You will not be able to change the type to which **Citreal** corresponds, as this involves recompiling all the Toolkit units (and you do not have their sources).

However, if you have, in addition, purchased the source of the Toolkit, and have a numeric coprocessor fitted to your machine, you can take advantage of the full range of Turbo Pascal real types. You may change the type to which `Citreal` corresponds by setting suitable conditional symbols (as described below), and fully recompiling the Toolkit (using the `Build` option). All the Toolkit units will then use the real type you specify, rather than Turbo Pascal's `real`; you do not have to make any further changes.

When you fully recompile the Toolkit with the numeric coprocessor off, `Citreal` defaults to the Turbo Pascal type `real`. If the coprocessor is on, then `Citreal` defaults to `double`. However, you can set it to any of the Turbo Pascal real types `real`, `single`, `double`, or `extended`, if you wish.

The following table gives the modulus of the smallest and largest values which the Turbo Pascal types can hold; note that this is *independent* of the sign of the value:

Type	Minimum value	Maximum value	Machine epsilon
`real`	1.0×10^{-38}*	1.5×19^{38}	1.9×10^{-12}
`single`	1.5×10^{-45}	3.4×10^{38}	6.0×10^{-8}
`double`	5.0×10^{-324}	1.7×10^{308}	1.2×10^{-16}
`extended`	3.4×10^{-4932}	1.1×10^{4932}	5.5×10^{-20}

Note that the Toolkit does not make available Turbo Pascal's `comp` type, which is not a floating-point type (merely an extended integer type).

The 'machine epsilon' mentioned above is the smallest value, ϵ, such that $1.0 + \epsilon \neq 1.0$ when evaluated by the machine. The Toolkit defines three constants, called `MINREAL`, `MAXREAL` and `EPSILON`, which take values as follows:

`MINREAL`	the minimum value a `Citreal` variable can hold
`MAXREAL`	the maximum value a `Citreal` variable can hold
`EPSILON`	the machine epsilon for a `Citreal` variable

When the `Citreal` type changes, these three constants must also change, if the Toolkit is to function properly. The Toolkit takes care of this automatically by using Turbo Pascal's conditional symbols to adjust all these values appropriately when you change the floating-point type you use.

---

* Note that, if the numeric coprocessor is switched on, the minimum value for the `real` type is about 2.9×10^{-39}.

To change the type defined to be `Citreal`, you must set a conditional symbol of the appropriate name. For example, to set `Citreal` to be identical to the Turbo Pascal type `Double`, you should define the conditional symbol `DOUBLE`. You can do this by following the drill below:

- Select the `Options` menu from the main menu (the easiest way to do this is to type `Alt`/`O`).
- Now select the `Compiler` option within this menu by pressing `C`.
- Select the `Conditional defines` option by pressing `C`.
- Now type in the name of the conditional symbol you want to define, pressing `Enter` when you have finished.

To make `Citreal` equivalent to a particular Turbo Pascal type, define one of the following conditional symbols:

Turbo Pascal type	Conditional symbol
`real`	`REAL`
`single`	`SINGLE`
`double`	`DOUBLE`
`extended`	`EXTENDED`

Note that selecting these types is only possible if you have a numeric coprocessor available, and switched on, when you are compiling your program.

When you have defined the appropriate conditional symbol, you will need to recompile those parts of the Toolkit your program uses. (As the type `Citreal` has changed, the Toolkit code which depends on `Citreal` must also be changed.) The easiest way to do this is to use the `Build` compiler option; load your main program (the one beginning with the `program` statement) into the editor, and type `Alt`/`C` followed by `B`.

The Toolkit defines two other useful constants, which are held in variables of type `Citreal`:

`LN10` $\log_e 10$ (to the full precision of a `Citreal`)

`MAXEXP` $\log_e$ `MAXREAL`

⟩ A1.2   Important types

Several important types, which are used throughout the Toolkit, are defined in `Cit_core`. However, they are used by units other than `Cit_core` itself, and have been fully described elsewhere in this book. Rather than describing them again ourselves, therefore, we give below the name of the unit which uses them, in whose description you can find full details.

Type	Used by unit
Expptr	Cit_eval
Textptr	Cit_text
Wimageptr	Cit_text
Citwindow	Cit_wind

⟩ A1.3   Useful numeric subprograms

Cit_core defines a number of useful numeric subprograms, which are listed very briefly in the table below:

Function	Returns
Float (i)	a Citreal equal to the integer i
Raise (x, y)	x^y
Sgn (x)	+1 if x is positive, −1 if it is negative, 0 if it is zero
Tan (x)	the tangent of x
Arccos (x)	the arc-cosine of x
Arcsin (x)	the arc-sine of x
Min (x, y)	the minimum of x and y
Max (x, y)	the maximum of x and y
Imax (i, j)	the maximum of the integers i and j
Imin (i, j)	the minimum of the integers i and j

These routines are used internally by the Toolkit, but are very useful in themselves; moreover, they are provided by other programming languages (notably FORTRAN). As the routines are well known, we merely give a description of their parameters, rather than a complete set of example programs!

⟩ *Function Float*

Routine name:	Float
Kept in unit:	Cit_core
Purpose:	Convert an integer into a Citreal

The unit name Cit_core must appear in the uses clause at the head of your program.

Signature:

```
function Float (i : integer) : Citreal;
```

Input parameters for Float:

Call: Float (ival)
ival: Integer value to be converted

Output parameters for Float:

Float returns a value of type Citreal, equal to its (integer) argument.

Error indicators for Float:

None.

) *Function Raise*

Routine name: **Raise**
Kept in unit: **Cit_core**
Purpose: Return first argument to the power of the second

The unit name **Cit_core** must appear in the uses clause at the head of your program.

Signature:

```
function Raise (value, power : Citreal) : Citreal;
```

Input parameters for Raise:

Call: Raise (val, pow)
val: Value to be raised
pow: Power by which to raise the value

Output parameters for Raise:

Raise returns val raised to the power pow.

Error indicators for Raise:

If **val** is zero, and **pow** is negative, **Raise** causes a "division by zero" error. If **val** is negative, and **pow** is not an integer, then **Raise** causes an "invalid floating-point operation" error.

) *Function Sgn*

Routine name: **Sgn**
Kept in unit: **Cit_core**
Purpose: Return the sign of the argument

The unit name **Cit_core** must appear in the uses clause at the head of your program.

Signature:

```
function Sgn (x : Citreal) : integer;
```

Input parameters for Sgn:

Call: **Sgn (val)**
val: Value whose sign is to be determined

Output parameters for Sgn:

Sgn returns +1 if **val** is positive, −1 if it is negative, and 0 if it is zero.

Error indicators for Raise:

None.

) *Function Tan*

Routine name: **Tan**
Kept in unit: **Cit_core**
Purpose: Return the tangent of the argument

The unit name **Cit_core** must appear in the uses clause at the head of your program.

Signature:

```
function Tan (x : Citreal) : Citreal;
```

Input parameters for Tan:

Call: **Tan (theta)**
theta: Angle whose tangent is to be calculated

Output parameters for Tan:

Tan returns the tangent of its argument.

Error indicators for Tan:

If **Tan** is called with an argument like $\pm\frac{\pi}{2}$, whose tangent is infinite, it returns a value of the correct sign with modulus **MAXREAL**.

⟩ *Function Arccos*

Routine name:	Arccos
Kept in unit:	Cit_core
Purpose:	Return the arc-cosine of the argument

The unit name **Cit_core** must appear in the uses clause at the head of your program.

Signature:

```
function Arccos (x : Citreal) : Citreal;
```

Input parameters for Arccos:

Call:	Arccos (val)
val:	Value whose arc-cosine is to be calculated

Output parameters for Arccos:

Arccos returns the arc-cosine of its argument.

Error indicators for Arccos:

None (if $|\text{val}| > 1$, then **Arccos** returns a value as though $|\text{val}| = 1$).

⟩ *Function Arcsin*

Routine name:	Arcsin
Kept in unit:	Cit_core
Purpose:	Return the arc-sine of the argument

The unit name **Cit_core** must appear in the uses clause at the head of your program.

Signature:

```
function Arcsin (x : Citreal) : Citreal;
```

Input parameters for Arcsin:

Call:	Arcsin (val)
val:	Value whose arc-sine is to be calculated

Output parameters for Arcsin:

Arcsin returns the arc-sine of its argument.

Error indicators for Arcsin:

None (if $|val| > 1$, then Arcsin returns a value as though $|val| = 1$).

) *Function Imin*

Routine name: Imin
Kept in unit: Cit_core
Purpose: Return the minimum of two integers

The unit name Cit_core must appear in the uses clause at the head of your program.

Signature:

```
function Imin (a, b : integer) : integer;
```

Input parameters for Imin:

Call: Imin (i, j)
i, j: Two integer values

Output parameters for Imin:

Imin returns the minimum of i and j.

Error indicators for Imin:

None.

) *Function Imax*

Routine name: Imax
Kept in unit: Cit_core
Purpose: Return the maximum of two integers

The unit name Cit_core must appear in the uses clause at the head of your program.

Signature:

```
function Imax (a, b : integer) : integer;
```

Input parameters for Imax:

Call: Imax (i, j)
i, j: Two integer values

Output parameters for Imax:
Imax returns the maximum of i and j.

Error indicators for Imax:
None.

) *Function Min*

Routine name: Min
Kept in unit: Cit_core
Purpose: Return the minimum of two Citreals

The unit name Cit_core must appear in the uses clause at the head of
your program.

Signature:

```
function Min (a, b : Citreal) : Citreal;
```

Input parameters for Min:
Call: Min (x, y)
x, y: Two values of type Citreal

Output parameters for Min:
Min returns the minimum of x and y.

Error indicators for Min:
None.

) *Function Max*

Routine name: Max
Kept in unit: Cit_core
Purpose: Return the maximum of two Citreals

The unit name Cit_core must appear in the uses clause at the head of
your program.

Signature:

```
function Max (a, b : Citreal) : Citreal;
```

Input parameters for Max:
Call: Max (x, y)
x, y: Two values of type Citreal

Output parameters for Max:

Max returns the maximum of x and y.

Error indicators for Max:

None.

⟩ **A1.4   Character constants**

The Toolkit uses internally a number of numeric constants, which are equal to the ordinal value of various characters. We give these below for completeness, rather than because we expect them to be useful!

Constant	Value	Constant	Value	Constant	Value
BELL	7	ESC	27	SPC	32
BRK	3	NINE	57	ZERO	48
BS	8	RTN	13		

The Toolkit, for its own internal purposes, also uses a number of named string constants. Once again, we give these for completeness:

Constant	Equivalent to string
SPACE	' '
PLUSINFINITY	'+∞'
MINUSINFINITY	'−∞'
INFINITY	'∞'
MATHPI	'π'

⟩ **A1.5   Colour constants**

Cit_core contains two types of colour constant. The first, and by far the most important, type is provided to give a pleasant interface to the Toolkit's screen-handling routines. All of these constants are of the type Citcolor, which is also defined in Cit_core. You will remember from Chapter 8 of this book that CITBLACK represents the colour black; the full list of colours so defined by the Toolkit is given below:

CITBLACK	black
CITBLUE	blue

`CITGREEN`	green
`CITCYAN`	cyan
`CITRED`	red
`CITMAGENTA`	magenta
`CITBROWN`	brown
`CITLIGHTGRAY`	light grey
`CITDARKGRAY`	dark grey
`CITLIGHTBLUE`	light blue
`CITLIGHTGREEN`	light green
`CITLIGHTCYAN`	light cyan
`CITLIGHTRED`	light red
`CITLIGHTMAGENTA`	light magenta
`CITYELLOW`	yellow
`CITWHITE`	white

Associated with these are the colours which define the default graphics colours within a window. These, too, are of the type `Citcolor`, and are listed below:

`AXISRED`	axes in red
`PIPYELLOW`	axis 'pips' in yellow
`LABLIGHTGRAY`	labels in light grey
`LINEWHITE`	lines in white

The second type of colour constant is a number, which is used internally by the Toolkit to inform the screen driving routines of the colour to be used. We list these constants below for completeness:

Constant	Represents	Value
`BLACK`	black	0
`BLUE`	blue	1
`GREEN`	green	2
`CYAN`	cyan	3
`RED`	red	4
`MAGENTA`	magenta	5
`BROWN`	brown	6
`LIGHTGRAY`	light grey	7
`DARKGRAY`	dark grey	8
`LIGHTBLUE`	light blue	9
`LIGHTGREEN`	light green	10
`LIGHTCYAN`	light cyan	11
`LIGHTRED`	light red	12
`LIGHTMAGENTA`	light magenta	13

Constant	Represents	Value
YELLOW	yellow	14
WHITE	white	15

The Toolkit also declares the global array `Citcolorset`, which contains the set of colours available to the Toolkit, and the constant `N_COLOR_MODES`. The type `Byte4` is also used internally by the Toolkit to represent screen coordinates.

) A1.6   Text primitives

We introduced text objects in Chapter 8. As we saw there, text objects are contained in variables of type `Textptr` (which is defined in `Cit_core`). Moreover, `Cit_core` contains the important subprograms: `New_text`, `Free_text` and `Shift_text`, which were introduced in Chapter 8.

) A1.7   Reference types

The Toolkit declares various types which enable you to reference arrays and variables by address. This is unfortunately necessary, given the restrictions in standard Pascal.

When you pass a variable to a Pascal subprogram, the type of the variable you pass must be identical to the type of the parameter. It is not enough for the two types to have the same properties; consider the following:

```
type one = array [1..10] of integer;
type two = array [1..10] of integer;
procedure p (parameter : one);
   ...
end;
```

Only variables of type **one** may be passed to the procedure **p**; not even variables of type **two** may be passed, even though both types are arrays of integers with exactly the same bounds.

This provides a great deal of type security, which is an important feature of the Pascal language; but it can, none the less, be a nuisance in numerical programming. Let us assume that we have written a small procedure, `zero_vector`, which takes a one-dimensional array of `Citreals` and sets all of them to zero. It would often be inconvenient if the variable passed to `zero_vector` had to be of a given type, rather than just any array of `Citreals` with (say) three components.

More importantly, consider a procedure, `invert`, which inverts a square matrix, represented by a two-dimensional array of `Citreals`. It would be extremely inconvenient to have to define an inversion procedure for every possible size of matrix: `invert2`, `invert3`, and so on. Since general methods exist for inverting a square matrix of any size, it would be much better if `invert` could be passed a matrix of any size, together with a parameter giving that size. This is impossible in strict Pascal, because arrays of different sizes are of different types.

However, Turbo Pascal allows us to obtain the *address* of an array, and pass it to a procedure. The size of the array does not matter, as the procedure expects an address (rather than an array), and is quite happy when an address is passed! The address of a Pascal object is of type `pointer`; if `pas_obj` is a Pascal object (either a variable or a subprogram), you can obtain its address by prefixing it with the operator `@`:

> `@ pas_obj`

(Alternatively, you may use the Turbo Pascal function `Addr` to return the address of an object:

> `Addr (pas_obj)`

is identical to `@ pas_obj`.)

Rather than use the type `pointer` throughout, the Toolkit defines various types which represent the addresses of particular objects. For example, the Toolkit type `Arrayptr` is used to store the addresses of arrays. Here is the full list:

Toolkit type	Holds address of
`Realptr`	`Citreal`
`Arrayptr`	an array of `Citreals`
`Coordptr`	a digitised shape
`Stringptr`	a Turbo Pascal `string`
`Funcptr`	a Pascal function or procedure

(In fact, `Arrayptr` is defined to be a pointer to the type `Datarray`, which is also declared in `Cit_core`; and `Coordptr` a pointer to the type `Coordarray`; but this need not concern most users!)

) **A1.8   Strings on the heap**

Pascal provides a *heap* on which data may be stored, independently of
any blocks and scoping. The type **Stringptr**, introduced in the previous
section, is used to hold strings on the heap, and has associated with it
two subprograms: **New_string**, which creates a new string on the heap,
and **Free_string**, which destroys a heap string, and returns the space it
occupied to the system. These subprograms are very similar to **New_text**
and **Free_text** which we introduced in Chapter 8; so we merely give the
briefest information about them here.

One important point: after you have released the storage occupied
by a heap string by calling **Free_string**, any attempt to use the heap
string's value will cause complete chaos (you might well crash the machine). So call **Free_string** only when you are quite sure that the string
value will not be used again.

) *Function New_string*

Routine name: **New_string**
Kept in unit: **Cit_core**
Purpose: Creates a string on the heap

The unit name **Cit_core** must appear in the uses clause at the head of
your program.

Signature:

```
function New_string (s : string) : Stringptr;
```

Input parameters for New_string:

Call: **New_string (s)**
s: String to be stored on the heap

Output parameters for New_string:

New_string returns the heap string it creates.

Error indicators for New_string:

None.

) *Procedure Free_string*

Routine name: **Free_string**

Kept in unit: `Cit_core`
Purpose: Destroys a string on the heap

The unit name `Cit_core` must appear in the uses clause at the head of your program.

Signature:

```
procedure Free_string (var p : Stringptr);
```

Input parameters for Free_string:

Call: `Free_string (p)`
p: Heap string to be destroyed

Output parameters for Free_string:

None.

Error indicators for Free_string:

None.

) A1.9   Dynamic arrays

Arrays in standard Pascal must have a size which is known at compile time. For example:

```
a1 : array [1..10] of Citreal;
```

is legal, because Pascal knows that `a1` must contain ten `Citreal` values. However,

```
a2 : array [1..n] of Citreal;
```

is illegal if n is a Pascal variable, even though its value is known when the declaration of `a2` is executed. This is a perfectly reasonable restriction, as arrays whose sizes are known only at run-time are very difficult for a compiler to cope with.

Languages (such as Ada) which allow declarations like those of `a2` above call arrays whose size is known only at run-time as *dynamic arrays*. Dynamic arrays would be a very useful facility for Pascal users; it is very convenient (say) to ask a user to input the number of data points on a graph, and then declare an array of exactly the right size to hold them.

The Toolkit provides facilities for declaring dynamic arrays, using the type **Arrayptr**, mentioned earlier, to hold them. It provides two subprograms: **New_array** creates space for a dynamic array on the heap; **Free_array** releases the space occupied by a dynamic array. These two subprograms work in much the same way as, say, the routines **New_text** and **Free_text**, which we met in Chapter 8.

New_array is passed the number of elements the array is to contain, and returns a heap array containing that number of **Citreal** variables. Turbo Pascal restricts the maximum size of an object on the heap to 65521 bytes; accordingly, the array returned by **New_array** may not contain more than 65521 DIV n **Citreals**, where n is the number of bytes required by a **Citreal**.

) *Function New_array*

Routine name:	**New_array**
Kept in unit:	**Cit_core**
Purpose:	Creates a dynamic array on the heap

The unit name **Cit_core** must appear in the uses clause at the head of your program.

Signature:

```
function New_array (n : word) : Arrayptr;
```

Input parameters for New_array:

Call:	**New_array (n)**
n:	Number of **Citreals** the array is to contain

Output parameters for New_array:

New_array returns the heap array it creates.

Error indicators for New_array:

New_array sets **Errorflag** if there is insufficient memory to hold the dynamic array, and returns NIL.

) *Procedure Free_array*

Routine name:	**Free_array**

Kept in unit: `Cit_core`
Purpose: Destroys a dynamic array on the heap

The unit name `Cit_core` must appear in the uses clause at the head of your program.

Signature:

```
procedure Free_array (var p : Arrayptr;
                          n : word);
```

Input parameters for Free_array:

Call: `Free_array (p)`
p: Dynamic array to be destroyed
n: Number of elements the dynamic array contains

Output parameters for Free_array:

p is set to `NIL`.

Error indicators for Free_array:

None.

) A1.10   Passing parameters which are subprograms

Standard Pascal allows you to pass procedures and functions as parameters to other procedures and functions. Turbo Pascal version 4.0 does not implement this, which is a pity; many of the Toolkit facilities (for instance, the evaluator, and many of the mathematical routines) depend on your being able to pass functions as parameters. This deficiency has been corrected by Turbo Pascal versions 5.0 and later, but we have designed the Toolkit to be used by people who only have access to Turbo Pascal version 4.0.

It is fairly straightforward to pass procedures and functions as parameters for use by the Toolkit; you merely precede their name by an `@`. For example,

   `@my_fun`

should be used to pass the subprogram `my_fun` to the Toolkit. In fact, the `@` causes Pascal to pass the *address* of the subprogram, rather than the subprogram itself, as the parameter. The type **Funcptr**, mentioned above, is compatible with subprograms passed in this way (such as **@my_fun**).

We know how to pass the subprograms as parameters: how do we call them, once they have been passed? The Toolkit provides, in **Cit_core**, facilities for doing this, in the form of the functions **Cpas0**, **Cpas1**, **Cpas2** and **Cpas3**. These are, in fact, just minor variations on one function; let us introduce **Cpas1** to begin with.

Cpas1 calls a Pascal function with *one* **Citreal** argument (this is where the 1 at the end of its name comes from), and returns its result (which must also be of type **Citreal**).

> **Cpas1 (@f, x);**

returns the same result as:

> **f (x);**

where **f** is a Pascal function of one variable of type **Citreal**.

(It is possible that you are still wondering why we should ever bother to use **Cpas1 (@f, x)**, when we could use **f (x)**. If this is the case, imagine that we have a variable, **func**, of type **Funcptr**, which holds the address of a Pascal function. If **func** holds **@f**, the address of **f**, then **Cpas1 (func, x)** will return **f (x)**, as we said above. But if **func** holds **@g** (where **g** is a different Pascal function with one **Citreal** parameter), then **Cpas1 (func, x)** will return **g (x)**. We could decide on which function to call by asking the user to decide the value of **func**; whereas, if we had written **f (x)**, it would only have been possible to call **f**.)

The other **Cpas** functions are defined in almost the same way: **Cpas0** returns the result of calling a function with no parameters; **Cpas2** returns the result of calling a function with two parameters; and **Cpas3** returns the result of calling a function with three.

) *Function Cpas0*

Routine name: **Cpas0**
Kept in unit: **Cit_core**
Purpose: Calls a Pascal function with no parameters, and returns the result

The unit name **Cit_core** must appear in the uses clause at the head of your program.

Signature:

```
function Cpas0 (f : Funcptr) : Citreal;
```

Input parameters for Cpas0:

Call:	Cpas0 (@f)
f:	Pascal function to call

Output parameters for Cpas0:

Cpas0 returns the result of calling f.

Error indicators for Cpas0:

None.

⟩ *Function Cpas1*

Routine name:	**Cpas1**
Kept in unit:	**Cit_core**
Purpose:	Calls a Pascal function with one parameter, and returns the result

The unit name **Cit_core** must appear in the uses clause at the head of your program.

Signature:

```
function Cpas1 (f : Funcptr;
                x : Citreal) : Citreal;
```

Input parameters for Cpas1:

Call:	Cpas1 (@f, x)
f:	Pascal function to call
x:	Argument passed to f

Output parameters for Cpas1:

Cpas1 returns the result of calling f (x).

Error indicators for Cpas1:

None.

⟩ *Function Cpas2*

Routine name: **Cpas2**
Kept in unit: **Cit_core**
Purpose: Calls a Pascal function with two parameters, and re-
                  turns the result

The unit name **Cit_core** must appear in the uses clause at the head of
your program.

Signature:

```
function Cpas2 (f : Funcptr;
                x : Citreal;
                y : Citreal) : Citreal;
```

Input parameters for Cpas2:

Call: **Cpas2 (@f, x, y)**
f: Pascal function to call
x, y: Arguments passed to **f**

Output parameters for Cpas2:

Cpas2 returns the result of calling **f** (**x, y**).

Error indicators for Cpas2:

None.

⟩ *Function Cpas3*

Routine name: **Cpas3**
Kept in unit: **Cit_core**
Purpose: Calls a Pascal function with three parameters, and
                  returns the result

The unit name **Cit_core** must appear in the uses clause at the head of
your program.

Signature:

```
function Cpas3 (f : Funcptr;
                x : Citreal;
                y : Citreal;
                z : Citreal) : Citreal;
```

Input parameters for Cpas3:

Call:	**Cpas3 (@f, x, y, z)**
f:	Pascal function to call
x, y, z:	Arguments passed to **f**

Output parameters for Cpas3:

Cpas3 returns the result of calling **f (x, y, z)**.

Error indicators for Cpas3:

None.

) A1.11   Linestrings

The longest line which can appear on a PC's screen (in any mode) contains 80 characters. To reflect this, the Toolkit defines a type **Linestring**, which is a Pascal string holding a maximum of 80 characters, and a constant, **LINESTRINGSIZE**, which is equal to 80. (This makes it easy for us to modify the Toolkit for machines which can display lines of a different length.) In addition, **Cit_core** declares a **Linestring** called **Blanks**, which contains merely spaces; this is used internally to pad output with spaces, but its length may vary unpredictably; please don't use it yourself!

) A1.12   Global error and keyboard variables

Cit_core contains the declarations of various global variables required throughout the Toolkit. The Boolean variables **Escape** and **Break** are used by the keyboard input routines to indicate whether $\boxed{\text{Esc}}$ and $\boxed{\text{Break}}$, respectively, have been pressed. Another Boolean, **Interruptflag**, is used to show whether the current Toolkit process has been interrupted from the keyboard.

Three variables are used for controlling the error behaviour of the Toolkit: the Boolean **Errorflag** is set **true** if any part of the Toolkit detects an error. If **Errorflag** is **true**, then the integer **Errorcode** will be set to a value reflecting the nature of the error, and the string **Errorstring** will contain a brief message (suitable for informing the user of what has happened). The error variables are set by the internal Toolkit procedure **Set_error**; this has two parameters, so that the call:

```
Set_error (errcode, errstr);
```

sets **Errorflag** to **true**, **Errorcode** to the integer **errcode**, and **Errorstring** to **errstr**. It is used throughout the Toolkit for reporting errors.

⟩ A1.13   The timer

Cit_core also declares a function **Timer**, which returns a value based on the time of day. This is used internally by the Toolkit to initialise its random-number generator.

⟩ *Function Timer*

Routine name: **Timer**
Kept in unit: **Cit_core**
Purpose: Return a value based on the time of day

The unit name **Cit_core** must appear in the uses clause at the head of your program.

Signature:

```
function Timer : longint;
```

Input parameters for Timer:
Call: **Timer**

Output parameters for Timer:
Timer returns a value of type **longint**, based on the time of day.

Error indicators for Timer:
None.

⟩ Appendix 2

⟩ Unit Cit_prim

This appendix describes the unit `Cit_prim`. This unit is similar to `Cit_core`, in that you are unlikely to use it directly yourself; but it is widely used throughout the Toolkit.

As `Cit_prim` is probably not a unit that you will use directly yourself (unless you are a sophisticated programmer with a particular use in mind), we have documented it fairly lightly. Those parts of the unit which appear to us to be more widely useful are described briefly, though not normally with comprehensive examples; the more obscure parts (like types used internally by the Toolkit) are mentioned by name only. In this way, we hope that you will be able to see the wood for the trees!

If you need full details of anything defined in `Cit_prim`, of course, you have only to read the source (which you can easily obtain if you wish).

⟩ A2.1   Screen modes

The Toolkit contains procedures which set the screen to either graphics or text mode. `Graphics_mode` puts the screen into graphics mode with immediate effect; `Text_mode` puts the screen into text mode with immediate effect.

Both of these procedures are parameterless; the call:

```
Graphics_mode;
```

puts the screen in graphics mode, and

```
Text_mode;
```

sets it to text mode. In both cases, the screen is completely cleared; the foreground text colour is set to light grey, and the background colour to black.

In addition, `Graphics_mode` sets the graphics colour to white; lines are to be drawn with full (rather than dotted or dashed) lines; and images are to be written directly onto the screen, rather than being modified by what is already there (the writemode is set to `ACTUAL_MODE`). The graphics port is set to the entire screen, and the palettes reset.

By default, `Graphics_mode` searches for the driver files it needs (the Turbo Pascal `BGI` files) in the current directory. If you wish `Graphics_mode` to search a different directory for these files, set the string variable `Driver_path` (which is defined in `Cit_prim`) to the name of the directory to be searched.

) *Procedure Text_mode*

Routine name: **Text_mode**
Kept in unit: **Cit_prim**
Purpose: Set the screen to text mode

The unit name `Cit_prim` must appear in the uses clause at the head of your program.

Signature:

```
procedure Text_mode;
```

Input parameters for Text_mode:
Call: **Text_mode;**

Output parameters for Text_mode:
None.

Error indicators for Text_mode:
None.

) *Procedure Graphics_mode*

Routine name: **Graphics_mode**
Kept in unit: **Cit_prim**
Purpose: Set the screen to graphics mode

The unit name `Cit_prim` must appear in the uses clause at the head of your program.

Signature:

```
procedure Graphics_mode;
```

Input parameters for Graphics_mode:
Call: `Graphics_mode;`

Output parameters for Graphics_mode:
None.

Error indicators for Graphics_mode:
None.

〉 *A2.1.1   Selecting a screen mode*

The procedure `Setup_mode` sets up the various system parameters for graphics mode (for instance, it works out which sort of graphics adapter is fitted). But it does *not* set graphics mode, or alter the screen; it merely 'prepares the ground' for a subsequent call of `Graphics_mode`. `Setup_mode` takes a parameter, the PC graphics mode to be used.

If its parameter is set to −1, `Setup_mode` chooses a mode which depends on the graphics adapter your computer is using. The resolution, number of colours, and number of the mode `Setup_mode` chooses are given in the following table:

Graphics adapter	Resolution	Number of colours	IBM mode number
CGA	640 × 200	2	6
MCGA	640 × 200	2	6
EGA	640 × 350	16	16
EGA64	640 × 350	4	16
VGA	640 × 350	16	16
EGAMono	640 × 350	2	15
Hercules	720 × 348	2	128

`Setup_mode` sets a number of integer variables defined in `Cit_prim`; let us first introduce the idea of text and graphics units and coordinates, so that we can understand their values.

Let us start with graphics coordinates. We could divide the screen into 'pixels', each of which could hold a 'dot' of some colour*. The position of any particular pixel could be given in terms of (x, y) coordinates; but here, we take the y-axis running down, rather than up, the screen, so the origin is at the top left of the screen. Coordinates like this, with the y-axis 'the wrong way round', are called *graphics coordinates*.

Now let us move on to text coordinates, which operate when the screen is displaying only characters (such as when it is running the Turbo Pascal compiler, or DOS). In this case, the coordinates are given by the row and column of a particular *character*, rather than a particular pixel; and the origin is at the *top* left of the screen. Thus characters on the top row of the screen have a y-coordinate of 1, and those on the bottom row have a y-coordinate of 25.

As a screen will display far fewer characters than it has pixels (because each character is made up of several pixels), the text and graphics coordinates of a particular point will be different. The following table gives the values of the variables which `Setup_mode` sets when it is called:

`N_screen_rows`	height of screen in text coordinates
`N_screen_cols`	width of screen in text coordinates
`N_screen_pages`	number of screen pages (see below)
`Screen_ht`	height of screen in graphics coordinates
`Screen_wd`	width of screen in graphics coordinates
`Cell_ht`	height of a text cell in graphics coordinates
`Cell_wd`	width of a text cell in graphics coordinates
`Char_ht`	height of a character in graphics coordinates
`Char_wd`	width of a character in graphics coordinates

For example, if the computer you are using is fitted with a CGA, the previous table tells us that `Setup_mode` will choose mode 6.

This has a graphics resolution of 640 × 200 pixels, so `Setup_mode` will set `Screen_ht` to 200, and `Screen_wd` to 640; it can display text in 25 rows of 80 columns, so `N_screen_rows` will be set to 25, and `N_screen_cols` to 80. A screen in this mode can display 16 colours in text mode, but only 2 in graphics mode; so `N_text_colors` will be set to 16, and `N_graphics_colors` to 2. A CGA screen in mode 6 can display 25 rows of 80 characters; so `Cell_ht` is set to 8 = 200/25, and `Cell_wd` is set to 8 = 640/80. The characters are held on an 8 × 8 bit-map; thus `Char_ht` and `Char_wd` are both 8.

---

* A 'pixel' is actually an acronym for 'picture element'.

A CGA has only one screen page, so **N_screen_pages** is set to 1 (for a discussion of screen pages, see the routines **Gp_visual_page** and **Gp_active_page**, described later in this appendix).

$\rangle$ *Procedure Setup_mode*

Routine name: **Setup_mode**
Kept in unit: **Cit_prim**
Purpose: Prepare parameters before setting graphics mode

The unit name **Cit_prim** must appear in the uses clause at the head of your program.

Signature:

```
procedure Setup_mode (ibmmode : integer);
```

Input parameters for Setup_mode:

Call: **Setup_mode (mode)**
mode: IBM graphics mode (if set to −1, **Setup_mode** decides the mode as described above)

Output parameters for Setup_mode:
None.

Error indicators for Setup_mode:
None.

$\rangle$ *A2.1.2   Clearing the screen*

An associated procedure, **Gp_clrscr**, is used by the Toolkit's graphics package to clear the screen.

$\rangle$ *Procedure Gp_clrscr*

Routine name: **Gp_clrscr**
Kept in unit: **Cit_prim**
Purpose: Clear the screen

The unit name **Cit_prim** must appear in the uses clause at the head of your program.

Signature:

```
procedure Gp_clrscr;
```

Input parameters for Gp_clrscr:

Call: `Gp_clrscr;`

Output parameters for Gp_clrscr:
None.

Error indicators for Gp_clrscr:
None.

⟩ *A2.1.3   Setting the viewport*

The lower levels of the Toolkit work in terms of *viewports* rather than windows. The procedure **Gp_viewport** sets the viewport, by defining two corners in graphics coordinates. The first corner (defined by (x_1, y_1)) is the upper left-hand corner of the viewport; the second (defined by (x_2, y_2)) is its lower right-hand corner. **Gp_viewport** is passed the graphics coordinates of both points, and an additional Boolean parameter called **clip**. If this is set to **true**, anything drawn which tries to extend outside the viewport will be clipped to fit it; otherwise, drawing will take place regardless of the presence of the viewport.

⟩ *Procedure Gp_viewport*

Routine name: **Gp_viewport**
Kept in unit: **Cit_prim**
Purpose: Set the size of the graphics viewport

The unit name **Cit_prim** must appear in the uses clause at the head of your program.

Signature:

```
procedure Gp_viewport (x1, y1 : integer;
                       x2, y2 : integer;
                       clip   : boolean;
```

Input parameters for Gp_viewport:

Call:	Gp_viewport (x1, y1, x2, y2, clip)
x1, y1:	Coordinates of the upper left-hand corner (in graphics units)
x2, y2:	Coordinates of the lower right-hand corner (in graphics units)
clip:	Boolean specifying whether lines extending outside the viewport are to be clipped

Output parameters for Gp_viewport:

None.

Error indicators for Gp_viewport:

None.

) *A2.1.4   Setting colours*

Cit_prim contains three integer variables which give the number of colours available. N_graphics_colors gives the number of colours available in graphics mode on your machine; N_text_colors gives the number of colours available in text mode; and N_screen_colors gives the number of colours available in the current screen mode (be it text or graphics).

The function Gp_convert_color is used internally by the Toolkit to produce a colour from a value of type Citcolor. The colour returned depends on the number of colours available in the current mode. If the colour you want is available in the current screen mode, then Gp_convert_color returns the code to produce that colour; otherwise, it does the best it can. For example, if you are using a screen mode in which green is available,

```
Gp_convert_color (CITGREEN);
```

will return the code for green; if the screen mode has only two colours available, Gp_convert_color will return the code for the foreground colour. This enables you to write programs which work sensibly regardless of the number of colours available. (To find out which colours Gp_convert_color returns in a particular screen mode, you will need to read its source; the algorithm is quite simple, but the detailed behaviour is too complex to describe here.)

) Function Gp_convert_color

Routine name: **Gp_convert_color**
Kept in unit: **Cit_prim**
Purpose: Convert **Citcolor** to suitable colour for the current mode

The unit name **Cit_prim** must appear in the uses clause at the head of your program.

Signature:

```
function Gp_convert_color (clr : Citcolor) : byte;
```

Input parameters for Gp_convert_color:
Call: **Gp_convert_color (clr)**
clr: **Citcolor** value giving the colour to be converted

Output parameters for Gp_convert_color:

Gp_convert_color returns an appropriate colour, based on its argument and the number of colours available in the current mode.

Error indicators for Gp_convert_color:
None.

) A2.1.5   Setting colours

Cit_prim contains three procedures for setting the colours for text, graphics, and background; all take a single parameter of type **Citcolor**. **Gp_text_color** sets the text foreground colour:

```
Gp_text_color (CITGREEN);
```

sets the text foreground colour to green.
    Gp_graph_color sets the current drawing colour in graphics mode:

```
Gp_graph_color (CITYELLOW);
```

would set the current drawing colour to yellow.
    Gp_back_color sets the current background colour, in either text or graphics mode:

```
Gp_back_color (CITBLUE);
```

would set the background colour to blue. Note that, in text mode, only certain colours may be used for the background colour (specifically black, blue, green, cyan, red, magenta, brown and light grey). If you try to use any of the other colours, the Toolkit will use one of the allowed colours instead.

⟩ *Procedure Gp_text_color*

Routine name: **Gp_text_color**
Kept in unit: **Cit_prim**
Purpose: Set text foreground colour

The unit name **Cit_prim** must appear in the uses clause at the head of your program.

Signature:

```
procedure Gp_text_color (clr : Citcolor);
```

Input parameters for Gp_text_color:
Call: **Gp_text_color (clr)**
clr: **Citcolor** value giving the colour to be used

Output parameters for Gp_text_color:
None.

Error indicators for Gp_text_color:
None.

⟩ *Procedure Gp_back_color*

Routine name: **Gp_back_color**
Kept in unit: **Cit_prim**
Purpose: Set background colour

The unit name **Cit_prim** must appear in the uses clause at the head of your program.

Signature:

```
procedure Gp_back_color (clr : Citcolor);
```

Input parameters for Gp_back_color:
Call: **Gp_back_color (clr)**
clr: **Citcolor** value giving the colour to be used

Output parameters for Gp_back_color:
None.

Error indicators for Gp_back_color:
None.

⟩ *Procedure Gp_graph_color*

Routine name: Gp_graph_color
Kept in unit: Cit_prim
Purpose: Set graphics drawing colour

The unit name **Cit_prim** must appear in the uses clause at the head of your program.

Signature:

```
procedure Gp_graph_color (clr : Citcolor);
```

Input parameters for Gp_graph_color:
Call: Gp_graph_color (clr)
clr: Citcolor value giving the colour to be used

Output parameters for Gp_graph_color:
None.

Error indicators for Gp_graph_color:
None.

⟩ **A2.2   Text in graphics mode**

The procedure **Gp_text_style** controls the size of, and font used for, the characters displayed *in graphics mode only*. The font is specified by a **word** variable, in just the same way as to the Turbo Pascal **SetTextStyle**; the size of the character is better thought of as a magnification (values from 1 to 10 are allowed). Text in graphics mode is written horizontally, and left-justified.

⟩ *Procedure Gp_text_style*

Routine name: Gp_text_style
Kept in unit: Cit_prim
Purpose: Set style and size of characters in graphics mode

The unit name `Cit_prim` must appear in the uses clause at the head of your program.

Signature:

```
procedure Gp_text_style (turbostyle, size : integer);
```

Input parameters for `Gp_text_style`:

Call: `Gp_text_style (charstyle, charsize);`
`charstyle`:   Font style (as passed to Turbo Pascal's `SetTextStyle`)
`charsize`:   Integer, giving the magnification required for text

Output parameters for `Gp_text_style`:
None.

Error indicators for `Gp_text_style`:
None.

) A2.3   Controlling the graphics writemode

The Toolkit procedure `Gp_writemode` allows you to control the way graphics you plot affect what is already on the screen. By default, anything you plot on the screen erases whatever is already there; `Gp_writemode` changes this default behaviour, according to its argument. This may be one of the following:

`ACTUAL_MODE`	plot overwrites what is already there
`INVERT_MODE`	inverse of plot overwrites what is already there
`EOR_MODE`	data plotted are exclusive-ORed with data already there
`OR_MODE`	data plotted are inclusive-ORed with data already there
`AND_MODE`	data plotted are logical-ANDed with data already there

For example, the call:

```
Gp_writemode (AND_MODE);
```

would mean that future plotted material would be ANDed with data already on the screen.

) *Procedure Gp_writemode*

Routine name: `Gp_writemode`
Kept in unit: `Cit_prim`
Purpose: Set graphics writemode

The unit name `Cit_prim` must appear in the uses clause at the head of your program.

Signature:

```
procedure Gp_writemode (writemode : byte);
```

Input parameters for `Gp_writemode`:
Call: `Gp_writemode (mode);`
mode: One of the values above

Output parameters for `Gp_writemode`:
None.

Error indicators for `Gp_writemode`:
None.

) **A2.4   Changing the line style**

The Toolkit procedure `Gp_line_style` allows you to change the way in which lines are plotted in graphics mode. By default, unbroken lines are joined between points where requested; `Gp_line_style` allows you to change this behaviour if you wish. `Gp_linestyle` takes one parameter, which may be one of the following:

```
SOLID_LINE     plot unbroken line
DASHED_LINE    plot dashed line
DOTTED_LINE    plot dotted line
```

For example, the call:

```
Gp_line_style (DASHED_LINE);
```

would result in future plots being made with dashed, rather than unbroken, lines.

) *Procedure Gp_line_style*

Routine name: `Gp_line_style`
Kept in unit: `Cit_prim`
Purpose: Set graphics line-drawing style

The unit name `Cit_prim` must appear in the uses clause at the head of your program.

Signature:

```
procedure Gp_line_style (ibmstyle : integer);
```

Input parameters for `Gp_line_style`:
Call: `Gp_line_style (style);`
`style`:   One of the values above

Output parameters for `Gp_line_style`:
None.

Error indicators for `Gp_line_style`:
None.

) **A2.5   Paging displays**

If your computer has an EGA, VGA or Hercules graphics adapter, it will have a number of screen *pages* associated with it. Only one of these is displayed at any one time, and only one can be written to by the graphics driver at any one time. The page which is currently being displayed is called the *visual* page; the page which is currently modifiable by graphics operations is called the *active* page.

Most importantly, the active and visual pages do *not* need to be the same. This can be very useful: consider a graph which takes some time to draw. It might be nice to display a message on the screen (such as "Please wait while I draw you a graph"), and be drawing the graph on another of the pages; then, when the graph is complete, change the visual page to be the one containing the graph, which will then appear in its complete form.

The Toolkit provides two procedures: `Gp_active_page`, which sets the active page, and `Gp_visual_page`, which sets the visual page. Those adapters which currently support multiple pages offer two pages, numbered zero and one. For example,

```
Gp_active_page (1);
```

sets the active page to page one. `Gp_active_page` and `Gp_visual_page` have no effect if your computer does not have an adapter which supports multiple graphics pages. The Toolkit sets the integer `N_screen_pages`, defined in `Cit_prim`, to the number of graphics pages available on your machine.

) *Procedure Gp_active_page*

Routine name: `Gp_active_page`
Kept in unit: `Cit_prim`
Purpose: Set active graphics page

The unit name `Cit_prim` must appear in the uses clause at the head of your program.

Signature:

```
procedure Gp_active_page (p : integer);
```

Input parameters for `Gp_active_page`:

Call: `Gp_active_page (page);`
page: Number of active page

Output parameters for `Gp_active_page`:
None.

Error indicators for `Gp_active_page`:
None (if the adapter only has one graphics page, `Gp_active_page` has no effect).

) *Procedure Gp_visual_page*

Routine name: `Gp_visual_page`
Kept in unit: `Cit_prim`
Purpose: Set visual graphics page

The unit name `Cit_prim` must appear in the uses clause at the head of your program.

Signature:

```
procedure Gp_visual_page (p : integer);
```

Input parameters for Gp_visual_page:

Call: `Gp_visual_page (page);`
page: Number of visual page

Output parameters for Gp_visual_page:

None.

Error indicators for Gp_visual_page:

None (if the adapter only has one graphics page, `Gp_visual_page` has no effect).

⟩ **A2.6 Filling and drawing**

The Toolkit provides extensive facilities for filling and drawing points, lines and shapes. The most flexible of these is called `Gp_fill`, which we shall introduce later; simpler routines, such as `Gp_line`, allow common operations to be performed much more simply.

 `Gp_pixel` colours a pixel at a particular point with the current graphics drawing colour. It is passed the (x, y) coordinates of the point *in graphics units*:

```
Gp_pixel (10, 20);
```

would colour the pixel at $(10, 20)$ with the current graphics drawing colour.

⟩ *Procedure Gp_pixel*

Routine name: `Gp_pixel`
Kept in unit: `Cit_prim`
Purpose: Colour a pixel

The unit name `Cit_prim` must appear in the uses clause at the head of your program.

Signature:

```
procedure Gp_pixel (x, y : integer);
```

Input parameters for Gp_pixel:

Call:	Gp_pixel (x, y);
x, y:	Coordinates of the point to colour

Output parameters for Gp_pixel:

None.

Error indicators for Gp_pixel:

None.

⟩ *A2.6.1   Drawing a line*

The Toolkit procedure `Gp_line` draws a line, in the current graphics drawing colour and style, between two points, specified by their (x, y) coordinates in graphics units. For example,

```
Gp_line (1, 10, 2, 20);
```

would draw a line between $(1, 10)$ and $(2, 20)$.

⟩ *Procedure Gp_line*

Routine name:	Gp_line
Kept in unit:	Cit_prim
Purpose:	Draw a line

The unit name `Cit_prim` must appear in the uses clause at the head of your program.

Signature:

```
procedure Gp_line (x1, y1, x2, y2 : integer);
```

Input parameters for Gp_line:

Call:	Gp_line (x1, y1, x2, y2);
x1, y1:	Coordinates of one end of the line
x2, y2:	Coordinates of other end of the line

Output parameters for Gp_line:

None.

Error indicators for Gp_line:

None.

⟩ *A2.6.2   Drawing a rectangle*

The procedure **Gp_rectangle** draws an outline rectangle, given the co-ordinates of two points in graphics units. The first is the upper left-hand corner of the rectangle, the second its lower right-hand corner. The outline is drawn with the current graphics drawing colour and line style. For example,

        Gp_rectangle (1, 10, 2, 20);

would draw an outline rectangle whose upper left-hand corner was at 1, 10, and whose lower right-hand corner was at 2, 20.

⟩ *Procedure Gp_rectangle*

Routine name: **Gp_rectangle**
Kept in unit: **Cit_prim**
Purpose: Draw an outline rectangle

The unit name **Cit_prim** must appear in the uses clause at the head of your program.

Signature:

        procedure Gp_rectangle (x1, y1, x2, y2 : integer);

Input parameters for Gp_rectangle:

Call: Gp_rectangle (x1, y1, x2, y2);
x1, y1:   Coordinates of upper left-hand corner of rectangle
x2, y2:   Coordinates of lower right-hand corner

Output parameters for Gp_rectangle:

None.

Error indicators for `Gp_rectangle`:
None.

⟩ *A2.6.3   Area filling*

Both `Gp_line` and `Gp_rectangle` use facilities provided by `Gp_fill`, which is a more general and powerful facility. `Gp_fill` fills a screen rectangle (given the same two corners as `Gp_rectangle`) with a specified colour, using a specified writemode. For example,

```
Gp_fill (x1, y1, x2, y2, CITGREEN, ACTUAL_MODE);
```

fills the rectangle with upper left-hand corner (x1, y1) and lower right-hand corner (x2, y2) with the colour green (using `ACTUAL_MODE` means that the original contents of the rectangle do not affect what is drawn).

⟩ *Procedure Gp_ fill*

Routine name: `Gp_fill`
Kept in unit: `Cit_prim`
Purpose: Fill an outline rectangle, with specified colour and writemode

The unit name `Cit_prim` must appear in the uses clause at the head of your program.

Signature:

```
procedure Gp_fill (x1, y1, x2, y2 : integer;
                   clr            : Citcolor;
                   writemode      : byte);
```

Input parameters for `Gp_fill`:

Call: `Gp_fill (x1, y1, x2, y2, colour, mode);`
x1, y1:   Coordinates of upper left-hand corner of rectangle
x2, y2:   Coordinates of lower right-hand corner
colour: Colour to use
mode: One of the values which may be passed to
             `Gp_writemode`

Output parameters for `Gp_fill`:
None.

Error indicators for Gp_fill:
None.

Finally, let us introduce the procedure **Gp_bar**, which draws a rectangle on the screen in the text *background* colour, thereby erasing everything it overwrites. Its parameters are identical to those of **Gp_rectangle**, but there is one vital difference: the coordinates of the corners are specified in *text*, not graphics, coordinates.

) *Procedure Gp_bar*

Routine name: **Gp_bar**
Kept in unit: **Cit_prim**
Purpose: Clear a rectangle on the screen

The unit name **Cit_prim** must appear in the uses clause at the head of your program.

Signature:

```
procedure Gp_bar (x1, y1, x2, y2 : integer);
```

Input parameters for Gp_bar:
Call: **Gp_bar (x1, y1, x2, y2);**
x1, y1:   Upper left-hand corner of area to be cleared
x2, y2:   Lower right-hand corner of area to be cleared

Output parameters for Gp_bar:
None.

Error indicators for Gp_bar:
None.

) **A2.7   Drawing polygons**

The Toolkit provides some routines for drawing, and filling, polygons with an arbitrary number of corners. **Gp_drawpoly** draws an outline polygon; **Gp_fillpoly** draws a filled-in polygon. Both are given two arguments: the number of vertices of the polygon, and a second argument of type **Coordptr**, which contains the coordinates of the points. **Coordptr** is defined in the unit **Cit_core**, so that programs which use **Gp_drawpoly** or **Gp_fillpoly** must include **Cit_core** as well as **Cit_prim** in their uses clause. **Coordptr** is, in fact, a pointer to a two-dimensional array of Turbo Pascal words; its full definition being:

```
type Coordptr = ^ Coordarray;
type Coordarray = array [1..16380, 1..2] of word;
```

Note that an array referenced by the type **Coordptr** can be smaller than a **Coordarray**; 16 380 should be regarded as a maximum for the first dimension. The simplest way to initialise a **Coordptr** is to use Turbo Pascal's typed constants:

```
const quad : array [1..5, 1..2] of word =
    (1, 1,  1, 100,  50, 100,  50, 1,  1, 1);
```

and then use the call:

```
Gp_drawpoly (5, @ quad);
```

which draws a quadrilateral on the screen (note that the last point must be identical to the first, in order to draw a closed polygon). The quadrilateral will have corners at graphics coordinates $(1, 1)$; $(1, 100)$; $(50, 100)$ and $(50, 1)$.

) *Procedure Gp_drawpoly*

Routine name:	**Gp_drawpoly**
Kept in unit:	**Cit_prim**
Purpose:	Draw an outline polygon

The unit name **Cit_prim** must appear in the uses clause at the head of your program.

Signature:

```
procedure Gp_drawpoly (n : integer;
                       p : Coordptr);
```

Input parameters for Gp_drawpoly:

Call:	**Gp_drawpoly (n, points);**
n:	Number of corners to draw
points:	Coordinates of the points

Output parameters for Gp_drawpoly:
None.

Error indicators for Gp_drawpoly:
None.

⟩ *Procedure Gp_ fillpoly*

Routine name: Gp_fillpoly
Kept in unit: Cit_prim
Purpose: Draw a filled polygon

The unit name Cit_prim must appear in the uses clause at the head of your program.

Signature:

```
procedure Gp_fillpoly (n : integer;
                       p : Coordptr);
```

Input parameters for Gp_fillpoly:
Call: Gp_fillpoly (n, points);
n: Number of corners to draw
points: Coordinates of the points

Output parameters for Gp_fillpoly:
None.

Error indicators for Gp_fillpoly:
None.

⟩ **A2.8   Saving and restoring images**

Cit_prim contains the low-level facilities required for saving and restoring portions of the screen. One example of their use is in pop-up windows, where a new window preserves the part of the screen over which it lies, and restores it when it is closed.

Cit_prim provides three subprograms to do this: Gp_getimage saves a screen image on the Pascal heap; Gp_putimage writes a saved image on the screen; and Gp_imagesize returns the amount of heap memory required to store an image. Images are saved in variables of type Gpimageptr, which is also defined in Cit_prim.

Note that images saved in this way are restricted in size. The reason for this is that images are saved on the Pascal heap, and the maximum size for an object on the heap is 65521 bytes. The amount of screen to which this corresponds depends on the graphics adapter fitted to your machine.

Suppose that you wish to save a screen image. Your first action must be to find the amount of memory required to hold it, by using **Gp_imagesize**. **Gp_imagesize** is passed the coordinates of two opposite corners of a rectangle (just as **Gp_rectangle** is), and returns the amount of memory required to contain it:

```
msize := Gp_imagesize (x1, y1, x2, y2);
```

would set the variable **msize** to the number of bytes required to hold the rectangle defined by the graphics coordinates $(x1, y1)$ and $(x2, y2)$. If the image is too large to be stored on the heap, **Gp_imagesize** returns zero.

You then need to initialise a variable of type **Gpimageptr** to a memory area of at least this size: the Turbo Pascal function **GetMem** is convenient for this. For example, if **gpi** is a variable of type **Gpimageptr**, then the call:

```
GetMem (gpi, msize);
```

would initialise **gpi** to the required amount of store.

) *Function Gp_imagesize*

Routine name:	**Gp_imagesize**
Kept in unit:	**Cit_prim**
Purpose:	Return amount of memory required to hold screen image

The unit name **Cit_prim** must appear in the uses clause at the head of your program.

Signature:

```
function Gp_imagesize (x1, y1 : integer;
                       x2, y2 : integer) : longint;
```

Input parameters for Gp_imagesize:

Call:	msize := Gp_imagesize (x1, y1, x2, y2);
x1, y1:	Coordinates of upper left-hand corner of rectangle
x2, y2:	Coordinates of lower right-hand corner

Output parameters for Gp_imagesize:

Gp_imagesize returns the number of bytes to hold the required screen image.

Error indicators for Gp_imagesize:

If the image would require too much memory to store, **Gp_imagesize** returns zero.

⟩ *A2.8.1 Saving an image*

The Toolkit routine **Gp_getimage** is used to save a portion of the screen into a variable of type **Gpimageptr** (which *must* have been initialised as described above). **Gp_getimage** is passed the coordinates of two opposite corners of a rectangle (just as **Gp_rectangle** is), and a variable into which the image is saved:

```
Gp_getimage (x1, y1, x2, y2, p);
```

saves the rectangle defined by the graphics coordinates (**x1,y1**) and (**x2,y2**), storing the result in **p**, which must have been initialised with the appropriate amount of memory.

⟩ *Procedure Gp_getimage*

Routine name: **Gp_getimage**
Kept in unit: **Cit_prim**
Purpose: Save a portion of the screen

The unit name **Cit_prim** must appear in the uses clause at the head of your program.

Signature:

```
procedure Gp_getimage (x1, y1, x2, y2 : integer;
                       b              : Gpimageptr);
```

Input parameters for Gp_getimage:

Call: **Gp_getimage (x1, y1, x2, y2, p);**
x1, y1:   Coordinates of upper left-hand corner of rectangle
x2, y2:   Coordinates of lower right-hand corner
p: Memory into which the image is saved

Output parameters for Gp_getimage:
The memory pointed to by p is modified to contain the screen image.

Error indicators for Gp_getimage:
None.

⟩ *A2.8.2   Restoring an image*

Finally, **Gp_putimage** may be used to restore the image onto the screen. It is passed the stored image, in a variable of type **Gpimageptr**; the graphics writemode to use; and the (x, y) coordinates at which the top left-hand corner of the image is to be placed. For example,

```
Gp_putimage (1, 10, gpi, ACTUAL_MODE);
```

would restore the image we saved with **Gp_getimage**, putting its upper left-hand corner at graphics coordinates $(1, 10)$, and using **ACTUAL_MODE** graphics drawing (that is, ignoring the present contents of that area of the screen).

⟩ *Procedure Gp_putimage*

Routine name:	**Gp_putimage**
Kept in unit:	**Cit_prim**
Purpose:	Save a portion of the screen

The unit name **Cit_prim** must appear in the uses clause at the head of your program.

Signature:

```
procedure Gp_putimage (x, y      : integer;
                       b         : Gpimageptr;
                       writemode : byte);
```

Input parameters for Gp_putimage:

Call:	Gp_putimage (x, y, p, mode);
x, y:	Coordinates of upper left-hand corner of rectangle
p:	Memory into which the image was saved
mode:	Graphics writemode to use

Output parameters for Gp_putimage:

None.

Error indicators for Gp_putimage:

None.

⟩ **A2.9   Writing text in graphics mode**

The procedure **Gp_graphic_text** writes a string, in the current graphics drawing colour, and the current writemode, at given graphics coordinates. For example:

```
Gp_graphic_text (1, 10, 'hello, world!');
```

would output the string **hello, world!** at the graphics coordinates $(1, 10)$.

⟩ *Procedure Gp_graphic_text*

Routine name: **Gp_graphic_text**
Kept in unit: **Cit_prim**
Purpose: Write a string at a given graphics position

The unit name **Cit_prim** must appear in the uses clause at the head of your program.

Signature:

```
procedure Gp_graphic_text (x, y : integer;
                           line : Linestring);
```

Input parameters for Gp_graphic_text:

Call: `Gp_graphic_text (x, y, s);`
x, y: Coordinates at which the string is to start
s: The string itself

Output parameters for Gp_graphic_text:

None.

Error indicators for Gp_graphic_text:

None.

⟩ A2.10   Controlling the keyboard

Cit_prim contains a number of procedures for low-level control of the keyboard and screen.

Before describing them, however, let us introduce two Boolean variables, **Escape** and **Break**, both of which are defined in Cit_core. These Booleans are set to **true** when the user presses Esc or Break, respectively. (In addition, **Break** is set to **true** if the user types Ctrl / C.) The Cit_prim routines both use and modify these variables; in this section, we shall describe the way these variables are changed.

The procedure Flush_keyboard_buffer flushes (deletes) any typed-ahead text in the keyboard buffer. This is useful, for example, when an error occurs, and you want to ask the user what to do (rather than read input which may have been typed ahead as program input).

```
Flush_keyboard_buffer;
```

flushes the keyboard buffer.

⟩ *Procedure Flush_keyboard_buffer*

Routine name: **Flush_keyboard_buffer**
Kept in unit: **Cit_prim**
Purpose: Flush any typed-ahead keyboard input

The unit name Cit_prim must appear in the uses clause at the head of your program.

Signature:

```
procedure Flush_keyboard_buffer;
```

Input parameters for Flush_keyboard_buffer:

Call: **Flush_keyboard_buffer;**

Output parameters for Flush_keyboard_buffer:

None. Both **Escape** and **Break** are set to **false**.

Error indicators for Flush_keyboard_buffer:
None.

⟩ *A2.10.1   Controlling the text cursor*

The two procedures **Text_cursor_off** and **Text_cursor_on** control whether the cursor is displayed when the screen is in text mode. By default, the cursor is switched on; but the call:

        Text_cursor_off;

will stop it being displayed. Calling **Text_cursor_on** reverses this effect.

⟩ *Procedure Text_cursor_off*

Routine name: **Text_cursor_off**
Kept in unit: **Cit_prim**
Purpose: Turn off the text cursor
The unit name **Cit_prim** must appear in the uses clause at the head of your program.

Signature:

        procedure Text_cursor_off;

Input parameters for Text_cursor_off:
Call: **Text_cursor_off**;

Output parameters for Text_cursor_off:
None.

Error indicators for Text_cursor_off:
None.

⟩ *Procedure Text_cursor_on*

Routine name: **Text_cursor_on**
Kept in unit: **Cit_prim**
Purpose: Turn on the text cursor
The unit name **Cit_prim** must appear in the uses clause at the head of your program.

Signature:

> procedure Text_cursor_on;

Input parameters for Text_cursor_on:

Call: Text_cursor_on;

Output parameters for Text_cursor_on:

None.

Error indicators for Text_cursor_on:

None.

) *A2.10.2   Quitting the program*

An associated procedure, Quit, puts the screen back into its last text mode, clears it, turns the cursor on, and exits from the program. This is the right way to 'tidy up' the screen after a program using graphics has finished.

) *Procedure Quit*

Routine name: Quit
Kept in unit: Cit_prim
Purpose: Clear the screen and leave

The unit name Cit_prim must appear in the uses clause at the head of your program.

Signature:

> procedure Quit;

Input parameters for Quit:

Call: Quit;

Output parameters for Quit:

None.

Error indicators for Quit:

None.

`Cit_prim` also contains a procedure `Gp_error`, which prints out an error message, after clearing the screen as for `Quit`, and then exits from the program.

) *A2.10.3 Reading individual keys*

`Cit_prim` contains two procedures, `Poll_keys` and `Read_keys`, for reading the keyboard. These are essentially identical, apart from one important difference: if no key is pressed at the time it is called, `Read_keys` will wait until one is, but `Poll_keys` will return a special value, indicating that no key is currently pressed. `Poll_keys` is useful, for example, if you want a program to run until the user presses a key.

Both `Read_keys` and `Poll_keys` are passed two arguments: an integer variable, which is normally set to the scan code of the key pressed, and a character variable, which is set to the character produced by the key (if one exists). The integer variable `Last_key`, which is declared in `Cit_prim`, is normally set to the ASCII code for the character read.

Consider the following call:

```
Read_keys (curcode, keychar);
```

where `curcode` is an integer variable, and `keychar` a character variable.

The paragraphs below give the various possible settings of `Last_key`, `curcode` and `keychar`, and how they are related to the key(s) the user actually pressed. The various settings may well seem complicated; but `Read_keys` and `Poll_keys` have been designed so that you can find out *exactly* which keys the user pressed. Remember that an IBM PS/2 keyboard has well over one hundred keys; and when Shift, Ctrl and Alt are considered as well, there are a great many possible key combinations!

- If the user has pressed a 'normal' key (for example, A) then `keychar` contains the corresponding character, `curcode` is set to zero, and `Last_key` set to the ASCII value of the character (65 for A).
- If `keychar` is equal to `#$FD`, then the user has pressed Alt and some other key. `curcode` is set to the extended scan code for the key combination; the global variable `Last_key` contains the ASCII value of the key. For example, if the user had pressed Alt/A, `keychar` would be `#$FD`, `curcode` equal to 30 (the extended scan code for Alt/A), and `Last_key` equal to 65 (the ASCII code for A).

- If **keychar** is equal to **#$FE**, then the user has pressed a function key. **curcode** is set to the number of the function key pressed, plus an offset. This offset is 0 if the function key was pressed on its own; 16 if [Shift] was pressed at the same time; 32 if [Ctrl] was pressed; and 48 if [Alt] was pressed. For example, if the user pressed [Shift]/[F6], then **keychar** would be set to **#$FE**, and **curcode** to 22 (= 16 + 6). **Last_key** would be set to 89 (the extended scan code for [Shift]/[F6]).

- If **keychar** is equal to **#$FF**, the user has pressed a cursor key, [Ins] or [Del]; **curcode** will normally contain the numerical value of the key (for example, 4 if [←] is pressed). If the user has pressed [Ins], **curcode** will equal 10; if [Del] is pressed, **curcode** will be set to 11. For example, if the user pressed the [↑] key, **keychar** would be **#$FF**; **curcode** would be 8 (the number printed on the 'up-arrow' key); and **Last_key** would be 72 (the extended scan code of [↑]).

- Otherwise, **keychar** is set to **$#FC**, **Last_key** to zero, and **curcode** to the extended scan code of the key pressed (for example, [Shift]/[Tab] sets **curcode** to 15).

Escape is set true if the user presses [Esc]; **Break** is set true if the user presses [Break] or [Ctrl]/[C].

If **Poll_keys** is called when no key is pressed, then **curcode** and **Last_key** are both set to zero, and **keychar** to **#0**. If **Read_keys** is called when no key is pressed, it waits until one is pressed, and returns the result.

) *Procedure Read_keys*

Routine name: **Read_keys**
Kept in unit: **Cit_prim**
Purpose: Read a key press

The unit name **Cit_prim** must appear in the uses clause at the head of your program.

Signature:

```
procedure Read_keys (var curcode : integer;
                     var keychar : char);
```

Input parameters for Read_keys:

Call: **Read_keys (code, ch);**

Output parameters for Read_keys:

code and ch are set as described above (for 'normal' keystrokes, code is zero, and ch set to the character read). For more details, and the settings of Escape, Break and Last_key, read the section above.

Error indicators for Read_keys:

None.

) *Procedure Poll_keys*

Routine name: Poll_keys
Kept in unit: Cit_prim
Purpose: Read a key press, if present

The unit name Cit_prim must appear in the uses clause at the head of your program.

Signature:

```
procedure Poll_keys (var curcode : integer;
                     var keychar : char);
```

Input parameters for Poll_keys:

Call: Poll_keys (code, ch);

Output parameters for Poll_keys:

code and ch are set as described above (for 'normal' keystrokes, code is zero, and ch set to the character read). If no key is pressed, ch is set to #0, and code to 0. For more details, and the settings of Escape, Break and Last_key, read the section above.

Error indicators for Poll_keys:

None.

A third procedure, Get_key_cursor, is almost identical to Read_keys. Whilst waiting for a key to be pressed, it flashes a cursor on the screen, which does not have to be in text mode.

) *Procedure Get_key_cursor*

Routine name: Get_key_cursor
Kept in unit: Cit_prim
Purpose: Read the keyboard, flashing cursor whilst waiting

The unit name **Cit_prim** must appear in the uses clause at the head of your program.

Signature:

```
procedure Get_key_cursor (var curcode : integer;
                          var keychar : char);
```

Input parameters for Get_key_cursor:

Call: `Get_key_cursor (code, ch);`

Output parameters for Get_key_cursor:

code and **ch** are set as described above (for 'normal' keystrokes, **code** is zero, and **ch** set to the character read). If no key is pressed, **ch** is set to **#0**, and **code** to 0. For more details, and the settings of **Escape**, **Break** and **Last_key**, read the section above.

Error indicators for Get_key_cursor:

None.

Another keyboard-related procedure is **Wait**, which waits a specified amount of time, or until the user presses Esc or Break, whichever is shorter. **Wait** takes a value of type **Citreal**, which is the expected time delay; thus:

```
Wait (10.3);
```

would wait 10.3 seconds, or until the user pressed Esc or Break. Notice that **Wait** can, in fact, delay the program longer than the time you specify, and so should not be used for accurate timing. This is particularly true for small time intervals; you should not rely on **Wait** to wait for much less than one second.

If the user presses Esc, **Wait** will set the variable **Escape** to **true**; if the user presses Break, **Wait** will set **Break** to **true**.

) *Procedure Wait*

Routine name: **Wait**
Kept in unit: **Cit_prim**
Purpose: Wait a specified amount of time, or until the user
                  interrupts

The unit name **Cit_prim** must appear in the uses clause at the head of your program.

Signature:

```
procedure Wait (secs : Citreal);
```

Input parameters for Wait:

Call: Wait (time);
time: Time to wait, in seconds

Output parameters for Wait:

None.

Error indicators for Wait:

If the user interrupts the wait with Esc or Break , either **Escape** or **Break** is set to **true**, accordingly.

) A2.11   Writing characters to the screen

Cit_prim provides two routines for writing characters to the screen, at given *text*, rather than graphics, coordinates. The procedure **Write_chars** writes characters from a string at the specified text (x, y) coordinates. It does *not* matter whether the screen is in text or graphics mode; the characters will be positioned as though the screen were in text mode.

The characters are drawn normally, with the exception that box characters (which are used for window margins) are drawn wider than normal; this is to improve the appearance of window margins.

```
Write_chars (1, 10, 'hello, there!');
```

writes the string **hello, there!** at text coordinates $(1, 10)$.

) Procedure Write_chars

Routine name: **Write_chars**
Kept in unit: **Cit_prim**
Purpose: Write a string, starting at given text coordinates

The unit name **Cit_prim** must appear in the uses clause at the head of your program.

Signature:

```
procedure Write_chars (x, y : integer;
                         str : Linestring);
```

Input parameters for Write_chars:

Call:	Write_chars (x, y, message);
x, y:	Text coordinates for the start of the string
message:	String to be written

Output parameters for Write_chars:
None.

Error indicators for Write_chars:
None.

The related procedure **Write_ch** writes the character with given ordinal value at the given text (x, y) coordinates. For example,

```
Write_ch (1, 10, 65);
```

would write the character with ordinal value 65 (**A**) at text coordinates $(1, 10)$.

⟩ *Procedure Write_ch*

Routine name:	Write_ch
Kept in unit:	Cit_prim
Purpose:	Write a character at given text coordinates

The unit name **Cit_prim** must appear in the uses clause at the head of your program.

Signature:

```
procedure Write_ch (x, y, ch : integer);
```

Input parameters for Write_ch:

Call:	Write_ch (x, y, c);
x, y:	Text coordinates for the start of the string
c:	Ordinal value of the character to be written

Output parameters for Write_ch:
None.

Error indicators for Write_ch:

None.

⟩ A2.12　Redefining characters

Define_user_char allows you to redefine characters whose ordinal value is greater than 127. It is passed the character to redefine, and eight byte values which represent the shape which the character is to produce.

Each byte contains eight bit values; as **Define_user_char** is passed eight byte values, the character is defined by an 8 × 8 matrix, as shown below:

128	64	32	16	8	4	2	1	
								10
								11
								12
								13
								14
								15
								16
								17

For example, the byte value **10** gives the bits which define the top row of the matrix. If a bit is set, that part of the character is displayed in the foreground colour; otherwise, in the background colour. If **10** has the value 3 (= 2 + 1), then the two right-hand bits on the first row of the matrix are set (with the others reset). This corresponds to the two right-hand pixels on the top row of the character being displayed in the foreground colour, with the others being displayed in the background colour.

Again, the byte value **17** affects the bits on the bottom row of the matrix; if **17** has the value 96 (= 64 + 32), then only the second and third bits from the left are set. In the bottom row of the character, therefore, only the pixels second and third from the left will appear in the foreground colour.

All the above may seem rather baffling; a simple way of getting used to defining characters is to try a few yourself; the bytes and bits turn out to be quite straightforward, once you have got used to them!

⟩ *Procedure Define_user_char*

Routine name: `Define_user_char`
Kept in unit: `Cit_prim`
Purpose: Redefine a character

The unit name `Cit_prim` must appear in the uses clause at the head of your program.

Signature:

```
procedure Define_user_char (chrcode : integer;
                            10, 11, 12, 13,
                            14, 15, 16, 17 : byte);
```

Input parameters for `Define_user_char`:

Call: `Define_user_char (ch, 10, 11, 12, 13,`
                                    `14, 15, 16, 17);`
ch: Character to redefine
10...17:   Bytes giving the definition (see above)

Output parameters for `Define_user_char`:
None.

Error indicators for `Define_user_char`:
None.

⟩ **A2.13   Miscellaneous**

The type `Citviewport` is defined and used internally by `Cit_prim`, as are the Booleans `Block_on`, `Turbo_graphics_on`, `Ibm_text_on`. The remaining internal variables are `Xcursor`, `Ycursor` (which contain cursor coordinates); `Xfrom`, `Yfrom` (which are used by `Cit_edit`); `Palette_2color`, `Palette_4color` (which contain the two- and four-colour palettes respectively); and `Turbo_graphport` (which contains the current viewport).

⟩ Appendix 3

⟩ Table of Error Messages

When the Toolkit code detects an error, it sets the global variable **Errorflag** to **true**. In addition, it sets the global variable **Errorcode** to an integer value which uniquely defines the error, and **Errorstring** to a brief description of the problem.

This appendix lists all the error codes and messages produced by the Toolkit, with a separate section for each unit.

Each error code consists of five digits. The first of these gives the Toolkit Level in which the error occurred: 0 for Level 0, 1 for Level 1, and so on. The next two digits give the serial number of the unit, and the last two the number of the error itself. Thus the error with code 10302 occurred in a Level 1 unit with serial number 3 (which is, in fact, **Cit_text**). The error code number within that unit was 2; by looking this up in this appendix, you can see that this corresponds to the message "Bad match to first line of file".

The error messages are deliberately terse, to save on memory space.

⟩ A3.1 Errors for unit Cit_core

00001:   Not enough room on heap for array

⟩ A3.2 Errors for unit Cit_prim

1010n: Turbo graphics error $-n$

⟩ A3.3 Errors for unit Cit_form

10201:   Bad format. Value does not fit in field
10202:   Bad format parameters

425

⟩ **A3.4**   **Errors for unit Cit_text**

10301:	Cannot open file
10302:	Bad match to first line of file
10303:	Cannot find file

⟩ **A3.5**   **Errors for unit Cit_eval**

10400:	Unexpected '.'
10401:	Digit expected in mantissa
10402:	Digit expected in exponent
10403:	Number out of range
10404:	Unexpected character in expression
10405:	')' expected
10406:	Number or identifier expected
10407:	End-of-expression expected
10408:	Unknown variable name
10409:	Subscript expected
10410:	Function argument expected
10411:	Unexpected subscript on constant
10412:	Unexpected subscript on simple variable
10413:	Constant expected
10414:	Division by zero
10415:	Square root of negative number
10416:	Result too large
10417:	Log of zero or negative number
10418:	Argument to **asin** or **acos** out of range
10419:	Fractional power of negative number
10420:	Negative power of zero
10421:	Subscript out of range
10422:	Argument to **rnd** out of range
10423:	No parse tree

⟩ **A3.6**   **Errors for unit Cit_wind**

10501:	Not enough room on heap for window image
10502:	Illegal character when reading integer
10503:	Illegal character when reading real

10504:	Bad window
10505:	Not enough room for file buffer
10506:	File error

〉 **A3.7   Errors for unit** `Cit_grap`

10601:	Not enough room for local array
10602:	Bad number of intervals
10603:	Bad axis range

〉 **A3.8   Errors for unit** `Cit_ctrl`

20701:	Not enough room for help messages
20702:	File error. Help file
20703:	No help available

〉 **A3.9   Errors for unit** `Cit_draw`

20801:	Not enough room in memory for local arrays
20802:	Not enough data
20803:	Bad range specified
20804:	Bad data specified
20805:	Invalid number of contours
20806:	Bad data array
20807:	Bad option specified
20808:	Escape. Sampled array not complete

〉 **A3.10   Errors for unit** `Cit_math`

20901:	Error in function evaluation
20902:	$f(a)$ and $f(b)$ not of opposite sign
20903:	Bad interval: high end $\leq$ low end
20904:	Spline needs ≥ 3 points
20905:	$x[i+1] \leq x[i]$ for some i
20906:	Must halve interval at least once
20907:	Not enough heap space

20908:   Tolerance must not be negative
20909:   Order must be ≥ 1
20910:   Order must be ≤ 30
20911:   Timestep must be positive
20912:   Minimum timestep must be positive
20913:   Power of 2 must be ≥ 1
20914:   No convergence after *maxiter* iterations

〉 A3.11   Errors for unit Cit_matr

21001:   Iteration for singular value failed
21002:   Accuracy not achieved after *maxiter* iterations in
              power method
21003:   Unsuitable initial value for eigenvalue
21004:   Accuracy not achieved after *maxiter* iterations in
              modified power method
21005:   Array size must be ≥ 1
21006:   Array size must be $\leq$ *maxn*
21007:   Accuracy must be > 0

〉 A3.12   Errors for unit Cit_disp

None.

〉 A3.13   Errors for unit Cit_menu

21201:   Option list empty

〉 A3.14   Errors for unit Cit_edit

None.

⟩ Index